Science Notebook

Glencoe Science

Earth Science

Geology, the Environment, and the Universe

Consultant
Douglas Fisher, PhD

Mc
Graw
Hill
Education

Cover (t) NASA, (c) T.J. Takahash/USGS, (b) Exactostock/Exactostock/SuperStock

MHEonline.com

Send all inquiries to:
McGraw-Hill Education
8787 Orion Place
Columbus, OH 43240

ISBN: 978-0-07-879203-8
MHID: 0-07-879203-7
Printed in the United States of America.

12 13 14 15 16 QVS 19 18 17 16 15

Table of Contents

Table of Contents

Using Your Science Notebook

This note-taking guide is designed to help you succeed in learning science content. Each chapter includes:

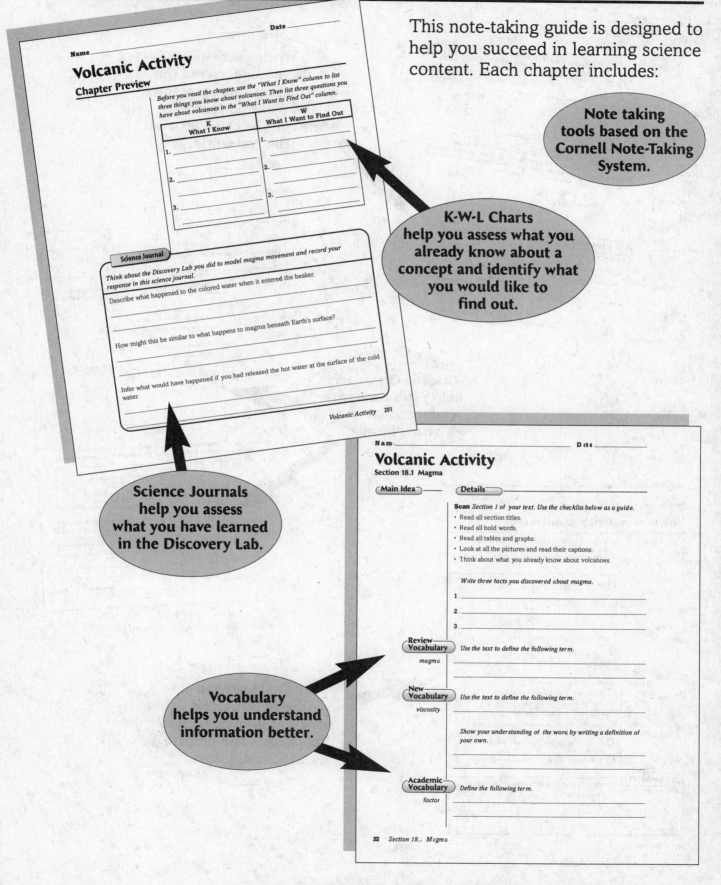

Note taking tools based on the Cornell Note-Taking System.

K-W-L Charts help you assess what you already know about a concept and identify what you would like to find out.

Science Journals help you assess what you have learned in the Discovery Lab.

Vocabulary helps you understand information better.

Volcanic Activity
Chapter Preview

Name _____ Date _____

Before you read the chapter, use the "What I Know" column to list three things you know about volcanoes. Then list three questions you have about volcanoes in the "What I Want to Find Out" column.

K What I Know	W What I Want to Find Out
1.	1.
2.	2.
3.	3.

Science Journal

Think about the Discovery Lab you did to model magma movement and record your response in this science journal.

Describe what happened to the colored water when it entered the beaker.

How might this be similar to what happens to magma beneath Earth's surface?

Infer what would have happened if you had released the hot water at the surface of the cold water.

Volcanic Activity 201

Volcanic Activity
Section 18.1 Magma

Name _____ Date _____

Main Idea — **Details**

Scan *Section 1 of your text. Use the checklist below as a guide.*
- Read all section titles.
- Read all bold words.
- Read all tables and graphs.
- Look at all the pictures and read their captions.
- Think about what you already know about volcanoes.

Write three facts you discovered about magma.

1.
2.
3.

Review Vocabulary — *Use the text to define the following term.*

magma

New Vocabulary — *Use the text to define the following term.*

viscosity

Show your understanding of the word by writing a definition of your own.

Academic Vocabulary — *Define the following term.*

factor

202   Section 18.1 Magma

Section 18.1 Magma (continued)

Name _____ Date _____

Main Idea | Details

How Magma Forms

Use with page 472.

List three factors that affect the formation of magma.

1. _____
2. _____
3. _____

Analyze Figure 18-1 to complete the table. Indicate in the table whether temperature, pressure, and melting point increase or decrease for each of the conditions.

	Pressure	Temperature	Melting Point of Wet Albite
Increasing Depth Below Earth's Surface			
Decreasing Depth Below Earth's Surface			

Illustrate the relationship between the melting point of a wet rock and a dry piece of the same rock under the same amount of pressure by sketching a thermometer and marking a possible temperature for each type of rock.

SYNTHESIZE If you were the engineer on an oil-drilling expedition explain how you might use a graph such as the one in Figure 18.1 in your text

Volcanic Activity 205

Writing activities help you understand the information being presented and make connections between the concepts and the real-world.

Section 18.2 Intrusive Activity (continued)

Name _____ Date _____

Main Idea | Details

Compare a sill and a dike. Place each characteristic below in the Venn diagram to show whether it is a characteristic of a sill, a dike, or both.

- cuts across preexisting rocks
- is parallel to the rocks it intrudes
- is a few centimeters to hundreds of meters thick
- many are coarse grained
- is a few centimeters to several meters wide
- is a pluton

Sill | Both | Dike

Plutons and Tectonics

Use with page 478.

Sequence the four steps involved in forming batholiths from mountain-building processes. The first step has been completed for you.

Two continental plates converge, forcing continental crust into the mantle.
OR
Two oceanic plates converge and one plate is subducted into the mantle.

Volcanic Activity 207

Graphic Organizers help you summarize information in a visual format.

Name _____ Date _____

Volcanic Activity Chapter Wrap-Up

In the "What I Wanted to Find Out" column, copy the questions you listed in the Chapter Preview. In the "What I Learned" column, write down the answers you discovered as you worked through the chapter.

K What I Wanted to Find Out	W What I Learned
1.	1.
2.	2.
3.	3.

Review

Use this checklist to help you study.

☐ Study your Science Notebook for this chapter.
☐ Study the definitions of vocabulary words.
☐ Review daily homework assignments.
☐ Reread the chapter and review the tables, graphs, and illustrations.
☐ Review the Section Assessment questions at the end of each section.
☐ Look over the Study Guide at the end of the chapter.

SUMMARIZE After reading this chapter, list three things you have learned about volcanic activity.

212 Chapter Wrap-Up

The Chapter Wrap-Up helps you assess what you have learned in the chapter and prepare for chapter tests.

---

vi *Using Your Science Notebook*

Note-Taking Tips

Your notes are a reminder of what you learned in class. Taking good notes can help you succeed in science. The following tips will help you take better classroom notes.

- Before class, ask what your teacher will be discussing in class. Review mentally what you already know about the concept.
- Be an active listener. Focus on what your teacher is saying. Listen for important concepts. Pay attention to words, examples, and/or diagrams you teacher emphasizes.
- Write your notes as clear and concise as possible. The following symbols and abbreviations may be helpful in your note-taking.

Word or Phrase	Symbol or Abbreviation	Word or Phrase	Symbol or Abbreviation
for example	e.g.	and	+
that is	i.e.	approximately	≈
with	w/	therefore	∴
without	w/o	versus	vs

- Use a symbol such as a star (★) or an asterisk (*) to emphasize important concepts. Place a question mark (?) next to anything that you do not understand.
- Ask questions and participate in class discussion.
- Draw and label pictures or diagrams to help clarify a concept.
- When working out an example, write what you are doing to solve the problem next to each step. Be sure to use your own words.
- Review you notes as soon as possible after class. During this time, organize and summarize new concepts and clarify misunderstandings.

Note-Taking Don'ts

- **Don't** write every word. Concentrate on the main ideas and concepts.
- **Don't** use someone else's notes as they may not make sense.
- **Don't** doodle. It distracts you from listening actively.
- **Don't** lose focus or you will become lost in your note-taking.

The Nature of Science
Chapter Preview

Before you read the chapter, use the "What I Know" column to list three things you know about science. Then list three questions you have about science in the "What I Want to Find Out" column.

K What I Know	W What I Want to Find Out
1. _____ _____	1. _____ _____
2. _____ _____	2. _____ _____
3. _____ _____	3. _____ _____

Science Journal

Describe at least three ways people use Earth science. You may find examples in the news, in the chapter, or think of examples in daily life.

_____

_____

_____

_____

_____

_____

_____

_____

The Nature of Science
Section 1.1 Earth Science

⟨Main Idea⟩ ————

⟨Details⟩ ————————————————

Skim *Section 1 of your text. Write three questions that come to mind from reading the headings and the illustration captions.*

1. _____

2. _____

3. _____

⟨New Vocabulary⟩ *Use your text to define each term.*

astronomy _____

meteorology _____

geology _____

_____

oceanography _____

environmental science _____

geosphere _____

hydrosphere _____

_____

atmosphere _____

biosphere _____

_____

Section 1.1 Earth Science (continued)

⟨**Main Idea**⟩ —— ⟨**Details**⟩ —————————————————

The Scope of Earth Science

Use with pages 6–7.

Organize *information about the topics each kind of scientist studies in the table below.*

Type of Scientist	Topics Studied
Astronomer	
Meteorologist	
Geologist	
Oceanographer	

Classify *the subspecialties of Earth Science by writing them under the appropriate areas of specialization. Use Table 1.1 to help you. You may write a subspecialty more than once.*

Earth Science

Meteorology	Geology	Oceanography	Astronomy

Identify *two recent events from the news that would be of interest to an Earth scientist, state what kind of scientist they would most interest, and classify the events under the appropriate subspecialties listed above.*

_____

_____

_____

_____

Section 1.1 Earth Science (continued)

⟨**Main Idea**⟩ ————	⟨**Details**⟩ ——————————————————
Earth's Systems *Use with pages 8–9.*	**Draw** *a diagram to help you understand and remember Earth's four main systems. Label each system.*
Earth Science in Your Everyday Life *Use with page 8–9.*	*List something from each of Earth's systems that humans use.* Lithosphere: _____ Hydrosphere: _____ Atmosphere: _____ Biosphere: _____

SYNTHESIZE

Identify one way that humans have changed each of Earth's systems.

Lithosphere: _____

Hydrosphere: _____

Atmosphere: _____

Biosphere: _____

The Nature of Science
Section 1.2 Methods of Scientists

Main Idea _____ **Details** _____

Skim *Section 2. Using what you already know and what you learn from skimming the section, describe how scientists use mathematics in their work.*

_____

_____

_____

_____

_____

New Vocabulary *Use your text to define each term.*

scientific methods _____

_____

hypothesis _____

_____

independent variable _____

dependent variable _____

_____

control _____

_____

Le Système Internationale d'Unités (SI) _____

_____

_____

scientific notation _____

Academic Vocabulary *Define the following term.*

bias _____

Section 1.2 Methods of Scientists (continued)

◯Main Idea◯ —————— **◯Details◯** —————————————————————————

The Nature of Scientific Investigations

Use with page 10–11.

Sequence *the steps of the scientific method by completing the flow chart below. Use Figure 1.6 to help you. Then draw arrows to show the order.*

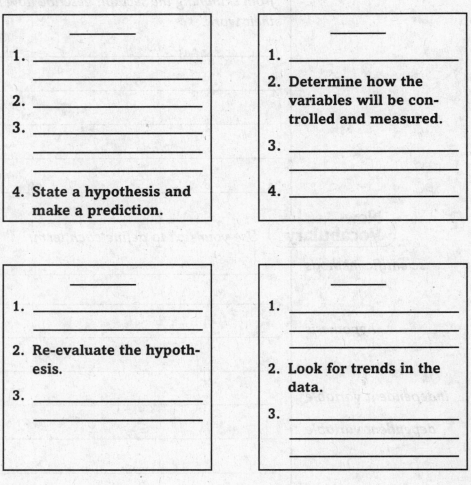

_____
1. _____
   _____
2. _____
3. _____
   _____
4. **State a hypothesis and make a prediction.**

_____
1. _____
2. **Determine how the variables will be controlled and measured.**
3. _____
   _____
4. _____

_____
1. _____
   _____
2. **Re-evaluate the hypothesis.**
3. _____
   _____

_____
1. _____
   _____
2. **Look for trends in the data.**
3. _____
   _____
   _____

Experimentation

Use with page 12.

Compare *independent and dependent variables by completing the following sentences.*

The _____ is a factor that is manipulated by the

experimenter. A(n) _____ is a factor that can change

if the _____ is changed.

Section 1.2 Methods of Scientists (continued)

| Main Idea | Details |

Safety in the Science Classroom

Use with page 954.

Summarize *the safety rules given in the Reference Handbook.*

Important Safety Rules	
1.	6.
2.	7.
3.	8.
4.	9.
5.	10.

Measurement

Use with page 14.

Organize *the concepts of measurement by completing the table.*

Quantity represented	What is measured	SI Unit	Abbreviation
	The amount of matter in an object	milliliter	mL
Length			m
			N
	Average vibrations of the particles that make up material	degrees Celsius	°C
	The amount of space occupied by an object	cubic meter	m³
Temperature		degrees Kelvin	K
	Amount of surface included within a set of boundaries		cm²
Time			S
	A measure of the amount of matter that occupies a given space		g/cm³

Section 1.2 Methods of Scientists (continued)

Main Idea ———   **Details** ————————————————————————

Area and Volume

Use with page 14.

Calculate *the area of area of the square and the volume of the cube.*

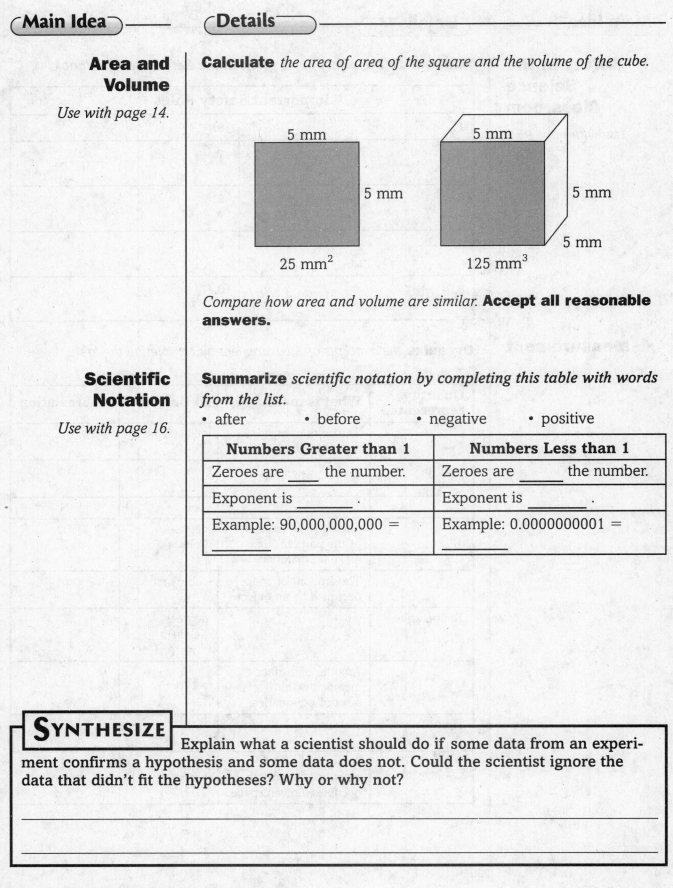

5 mm

5 mm

25 mm^2

5 mm

5 mm

5 mm

125 mm^3

Compare how area and volume are similar. **Accept all reasonable answers.**

Scientific Notation

Use with page 16.

Summarize *scientific notation by completing this table with words from the list.*

- after   • before   • negative   • positive

Numbers Greater than 1	Numbers Less than 1
Zeroes are _____ the number.	Zeroes are _____ the number.
Exponent is _____ .	Exponent is _____ .
Example: 90,000,000,000 = _____	Example: 0.0000000001 = _____

SYNTHESIZE Explain what a scientist should do if some data from an experiment confirms a hypothesis and some data does not. Could the scientist ignore the data that didn't fit the hypotheses? Why or why not?

_____

_____

The Nature of Science
Section 1.3 Communication in Science

⟨Main Idea⟩————— **⟨Details⟩**———————————————————

Scan *Section 3 of your text. Use the checklist as a guide.*

• Read all section titles.

• Read all bold words.

• Read all tables and graphs.

• Look at all the pictures and read their captions.

• Think about what you already know about communicating in science.

Write three facts you discovered about how scientists communicate.

1. _____

2. _____

3. _____

⟨New Vocabulary⟩

scientific model

scientific theory

scientific law

Write a paragraph that explains the difference between a theory and a law in science.

_____

_____

_____

_____

_____

_____

Section 1.3 Communication in Science (continued)

Main Idea —   **Details** _____

Communicating Results

Use with pages 17–18.

Create *an outline of the information under the heading "Communicating Results."*

  I. Communicate data and results to others so they can:

    A. _____

    B. _____

    C. _____

  II. Ways scientific results are communicated

    A. _____

    B. _____

Identify *the parts of this graph.*

1. Highlight the title of the graph below.

2. Label the axis where the dependent variable is plotted "Dependent."

3. Label the axis where the independent variable is plotted "Independent."

4. Circle the labels that tell what units each variable is expressed in.

_____

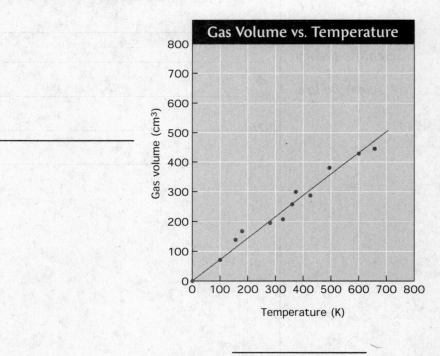

Section 1.3 Communication in Science (continued)

Main Idea	Details

Models

Use with page 18.

Summarize *the information under the heading "Models" by completing the sentences below.*

1. A scientific model is _____

_____.

2. A model should be _____.

3. Models can change when _____.

Theories and Laws

Use with page 19.

Compare *and contrast theories and laws. Place each statement in the Venn diagram to show whether it is true for theories, laws, or both.*

- Based on observations
- Can change with the discovery of new data
- Describes a natural phenomenon
- May not explain what it describes
- Used to explain scientific laws

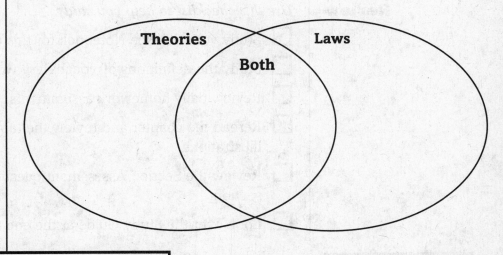

Theories   Laws

Both

┌──┐
│ **REAL-WORLD CONNECTION** |
│ Suppose you are a doctor who has dis-│
│ covered a cure for a fatal disease. What method would you use to communicate your │
│ discovery to the world? Why would you choose that method? │
│ │
│ _____ │
│ │
│ _____ │
└──┘

The Nature of Science Chapter Wrap-Up

In the "What I Wanted to Find Out" column, copy the questions you listed in the Chapter Preview. In the "What I Learned" column, write down the answers you discovered as you worked through the chapter.

W What I Wanted to Find Out	L What I Learned
1. _____ _____	1. _____ _____
2. _____ _____	2. _____ _____
3. _____ _____	3. _____ _____

Review *Use this checklist to help you study.*

☐ Study your Science Notebook for this chapter.

☐ Study the definitions of vocabulary words.

☐ Review daily homework assignments.

☐ Reread the chapter and review the tables, graphs, and illustrations.

☐ Review the Section Assessment questions at the end of each section.

☐ Look over the Study Guide at the end of the chapter.

SUMMARIZE

below. Write the main idea of each section of this chapter on the lines

1. _____

2. _____

3. _____

Mapping Our World
Chapter Preview

Before you read the chapter, use the "What I Know" column to list three things you know about maps. Then list three questions you have about maps in the "What I Want to Find Out" column.

K What I Know	W What I Want to Find Out
1. _____ _____	1. _____ _____
2. _____ _____	2. _____ _____
3. _____ _____	3. _____ _____

Science Journal

Think about the Launch Lab you completed to make and use a map and record your response in this science journal. What kind of information did you include in your map? What landmarks were omitted that would have been helpful? Explain your answer.

_____

_____

_____

_____

_____

_____

Mapping Our World
Section 2.1 Latitude and Longitude

Main Idea ———————————— **Details** ———————————————————————

Skim *Section 1 of your text. Write three questions that come to mind from reading the headings and the illustration captions.*

1. _____

2. _____

3. _____

New Vocabulary   *Use your text to define the following term.*

cartography   _____

_____

Use the terms to label the figure below.

equator

latitude

longitude

prime meridian

International Date Line

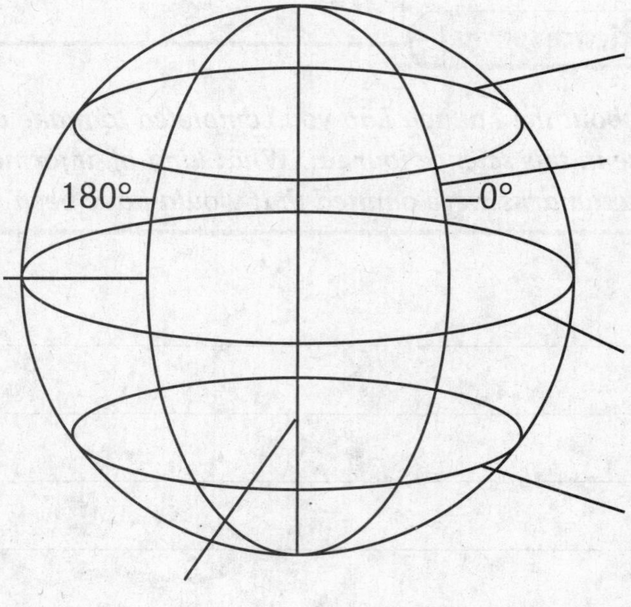

Section 2.1 Latitude and Longitude (continued)

Main Idea ——— Details ———

Latitude

Use with pages 30–31.

Organize *information about latitude by completing this graphic organizer. Use Figure 2.1 and other information in your text.*

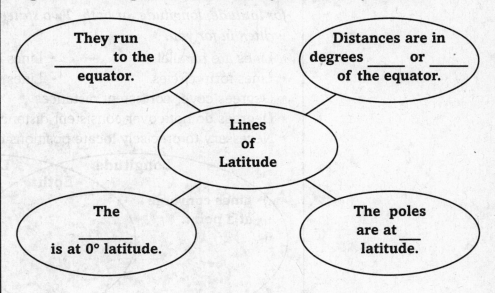

They run _____ to the equator.

Distances are in degrees _____ or _____ of the equator.

Lines of Latitude

The _____ is at 0° latitude.

The poles are at ___ latitude.

Summarize *the information about degrees of latitude by completing the table.*

Degrees of Latitude

	Symbol	Actual Distance
Degree		
	'	

Longitude

Use with pages 31–33.

Complete *each sentence to help you remember information about longitude.*

Lines of longitude are also known as _____ . Longitude is

distance in degrees ____ or ____ of the prime meridian. The

_____ is the reference point for meridians. It represents _°

longitude. The prime meridian goes through _____ .

Points west of the prime meridian are numbered from _° to ___°

west longitude. Points east of the prime meridian are numbered

from 0° to 180° _____ .

Section 2.1 Latitude and Longitude (continued)

⟨**Main Idea**⟩ ———   ⟨**Details**⟩ ———————————————————

Use with pages 30–31.

Compare and contrast *information about latitude and longitude. Place each statement in the Venn diagram to show whether it is true for latitude, longitude, or both. Two statements have already been written in for you.*

- Lines are parallel.
- Lines form circles.
- Degrees cover consistent distances.
- Degrees do not cover consistent distances.
- Necessary to precisely locate positions on Earth

- Lines converge at a point.
- Lines form semicircles.

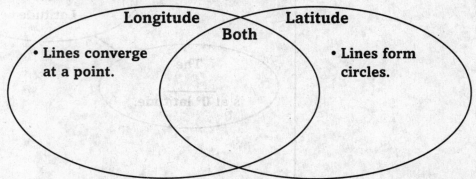

Longitude   Both   Latitude

• **Lines converge at a point.**

• **Lines form circles.**

Time Zones

Use with pages 32–33.

Analyze *the figure below showing U.S. time zones. Assume that it is 9:00 in the Mountain time zone. Draw the hands on the other clocks to show the time it would be in each of the other U.S. time zones.*

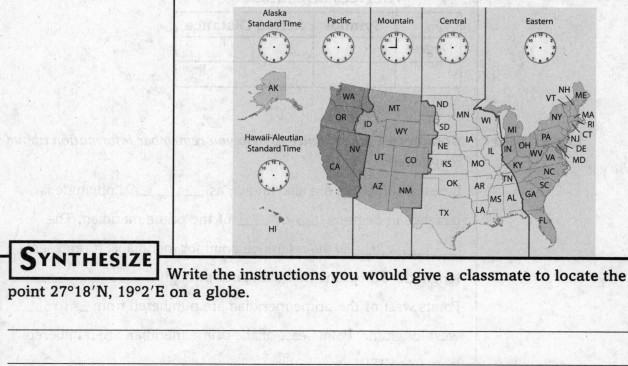

SYNTHESIZE
Write the instructions you would give a classmate to locate the point 27°18′N, 19°2′E on a globe.

_____

_____

Mapping Our World
Section 2.2 Types of Maps

Main Idea ———   **Details** ————————————

Skim *Section 2 of your text. List the three map projections mentioned in the headings and illustration captions.*

1. _____

2. _____

3. _____

New Vocabulary   *In the left column, write the terms defined below.*

_____   *a map that has parallel lines of latitude and longitude*

_____   *a map made by projecting points and lines from a globe onto a cone*

_____   *a map made by projecting points and lines onto a piece of paper that touches a globe at a single point*

_____   *a map that shows changes in elevation of Earth's surface*

_____   *a line on a map that connects points of equal elevation*

_____   *the difference in elevation between two side-by-side contour lines on a map*

_____   *a table that explains what the symbols on a map represent*

_____   *the ratio between distances on a map and actual distances on the surface of Earth*

Academic Vocabulary   *Define the following term.*

ratio   _____

_____

Section 2.2 Types of Maps (continued)

Main Idea ———— **Details** ————————————————————

Mercator Projections

Use with page 34.

Complete *the table to organize information about Mercator projections.*

Mercator Projections

Made by	
Advantages	
Disadvantages	
Uses	

Conic Projections

Use with page 35.

Organize *information about conic projections by filling in the graphic organizer.*

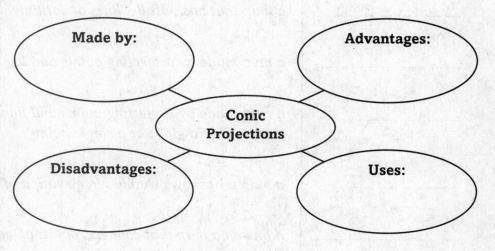

Gnomonic Projections

Use with page 35.

Complete *the following sentences to help you understand gnomonic projections.*

Gnomonic projections are made by _____

_____

The advantage of these projections _____

_____

The disadvantage of gnomonic projections is _____

_____

They are used by _____

Section 2.2 Types of Maps (continued)

Main Idea	Details

Topographic Maps

Use with pages 36–37.

Assess *your understanding of topographic maps. Write two questions that might appear on a quiz. Then write the answers.*

Question: Why can contour lines on a map never cross?_____

Answer:_____

_____

_____

Question:_____

Answer:_____

_____

Map Legends

Use with page 39.

Create *symbols for a map legend in the spaces below. Include symbols for four features in your house, school, or neighborhood.*

Map Scales

Use with pages 39–40.

Summarize *information about map scales by completing the table.*

Type of Scale	Expressed as	Example
Verbal scale		
Graphic scale		
Fractional scale		

REAL-WORLD CONNECTION

Explain *which type of map you would probably find in each of the following places and why: a car; a hiker's backpack; a sailboat.*

_____

_____

_____

Mapping Our World
Section 2.3 Remote Sensing

```
╭ Main Idea ╮ ───    ╭ Details ╮ ─────────────────────────
```

Predict *what you will learn about in Section 3. Read the title of the section and the first paragraph after the section title. Then write what you think this section will be about.*

_____

_____

_____

_____

```
╭ New Vocabulary ╮
```
Use your text to define each term.

remote sensing _____

_____

Landsat satellite _____

_____

Topex/Poseidon satellite _____

_____

Global positioning system _____

_____

sonar _____

_____

Geographic Information system _____

_____

```
╭ Academic Vocabulary ╮
```
Define the following term.

comprehensive _____

_____

Section 2.3 Remote Sensing (continued)

Main Idea	Details
Landsat Satellite *Use with page 41.*	*How do Landsat Satellites work?* _____ _____ *Devise an experiment using a Landsat satellite and hypothesize results.* _____ _____ _____ _____ _____ _____ _____ _____ _____ _____
Topex/Poseidon Satellite *Use with page 42.*	**Model** *a Topex/Poseidon satellite by drawing a diagram. Use Figure 2.13 in your text for reference. Draw and label your diagram showing how these satellites map the ocean floor.*

Section 2.3 Remote Sensing (continued)

⊂Main Idea⊃ ———— ⊂Details⊃ ——————————————————————

Landsat and Topex/Poseidon Satellites The Global Positioning System and Sea Beam

Use with pages 41-45.

Distinguish *between the remote sensing methods discussed in this chapter by completing the table below.*

Type of Satellite	Kind of Waves	Where Waves are Emitted
Landsat satellite		
TOPEX/POSEIDEN satellite		
Global Positioning System		
Sea Beam		

Organize *information about Sea Beam technology by completing the graphic organizer below.*

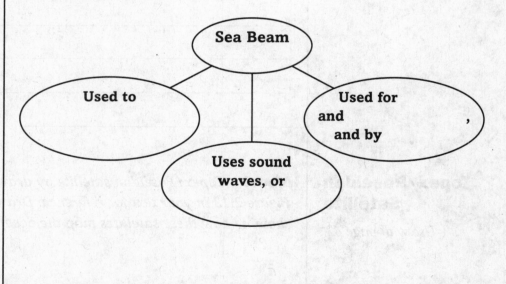

SYNTHESIZE *Predict which satellite mentioned in this section will be used the most in the future. Describe how your chosen satellite works and compare its advantages and disadvantages with other satellites to explain your answer.*

_____

_____

Tie-It-All-Together

SYNTHESIZE

Think about what you have learned in this chapter about maps.
In the space below, design a map of your classroom.
Create a map legend to represent lights, electrical outlets, books, and other objects.
Try to make your map as accurate as possible.
Be sure to include a scale.

Mapping Our World Chapter Wrap-Up

In the "What I Wanted to Find Out" column, copy the questions you listed in the Chapter Preview. In the "What I Learned" column, write down the answers you discovered as you worked through the chapter. Use this checklist to help you study.

W What I Wanted to Find Out	L What I Learned
1. _____ _____	1. _____ _____
2. _____ _____	2. _____ _____
3. _____ _____	3. _____ _____

Review *Use this checklist to help you study.*

☐ Study your Science Notebook for this chapter.

☐ Study the definitions of vocabulary words.

☐ Review daily homework assignments.

☐ Reread the chapter and review the tables, graphs, and illustrations.

☐ Review the Section Assessment questions at the end of each section.

☐ Look over the Study Guide at the end of the chapter.

SUMMARIZE

After reading this chapter, list three things you have learned about maps.

_____

_____

_____

Matter and Change

Chapter Preview

Before you read the chapter, use the "What I Know" column to list three things you know about matter and atomic structure. Then list three questions you have about matter and atomic structures in the "What I Want to Find Out" column.

K What I Know	W What I Want to Find Out
1. _____ _____	1. _____ _____
2. _____ _____	2. _____ _____
3. _____ _____	3. _____ _____

Science Journal

Think about the Launch Lab you did to identify one substance added to breakfast cereals and record your response in this science journal.

Describe what you see on the end of the pencil stirrer. Study the cereal box to determine what the substance on the end of the magnet might be.

_____

_____

_____

_____

_____

_____

_____

Matter and Change
Section 3.1 Matter

(Main Idea)	(Details)

Scan *Section 1 of your text. Use the checklist below as a guide.*

- Read all headings.
- Read all bold words.
- Read all tables and graphs.
- Look at all the pictures and read their captions.
- Think about what you already know about elements.

Academic Vocabulary   *Write a fact about atoms.*

_____

New Vocabulary   *In the left margin, write the terms defined below.*

_____   *anything that has volume and mass*

_____   *a substance that cannot be broken down into simpler substances by physical or chemical means*

_____   *the center of an atom, made up of protons and neutrons*

_____   *a particle that has mass and a positive electrical charge*

_____   *a particle with about the same mass as a proton but with no electrical charge*

_____   *a particle with little mass and a negative electric charge*

_____   *the number of protons in an atom's nucleus*

_____   *the combined number of protons and neutrons in an atom*

_____   *atoms of the same element with different mass numbers*

_____   *an atom that gains or loses an electron with a net electric charge*

Academic Vocabulary   *Define the following term.*

region   _____

_____

Section 3.1 Matter (continued)

⟨**Main Idea**⟩ ——— ⟨**Details**⟩ ————————————————————

Elements

Use with pages 60–61.

Describe *elements by completing the prompts below.*

An element is a _____

_____ . *There are* __

elements that occur naturally in the universe. A chemical symbol is

a _____ .

_____ .

Atoms

Use with pages 60–61.

Organize *information about the structure of atoms by writing the letter for each statement in the correct area of the diagram. Some letters will be used more than once.*

a. consists of protons, neutrons, and electrons

b. determines the atomic number

c. has exactly the same number of protons and electrons

d. has little mass

e. has mass

f. has negative charge

g. has no charge

h. has positive charge

i. helps determine the mass number

j. is found in an energy level around the nucleus

k. is made up of protons and neutrons

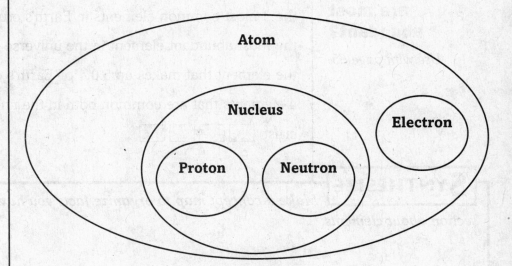

Section 3.1 Matter (continued)

Main Idea	Details

Electrons in Energy Levels

Use with pages 63–64.

Draw *the figures below to help you understand isotopes. Use information from your book to help you.*

1. Draw a chlorine-35 atom.	1. Draw a chlorine-37 atom.
2. Label the nucleus with the number of protons and neutrons inside.	2. Label the nucleus with the number of protons and neutrons inside.
3. Circle the valence energy level.	3. Circle the valence energy level.

Isotopes

Use with page 62.

Name the relationship of the two atoms you drew above.

_____

State if the two atoms would have the same properties, or not.

_____

What elements are most abundant?

Use with page 65.

Analyze *Figure 3.6 in your text. Write the chemical symbols for elements that match the following descriptions.*

the 3 most common elements in Earth's crust ☐ ☐ ☐

the most abundant element in the universe ☐

the element that makes up 5.0% of Earth's crust ☐

4 elements that are common both in the universe and in the earth's

crust ☐ ☐ ☐ ☐

SYNTHESIZE

Make a concept map to organize facts you have learned in this section about elements.

Matter and Change
Section 3.2 Combining Matter

Main Idea —— **Details** ————————————————

Scan *the objectives and the main headings for Section 2 of your text. Write three questions that come to mind.*

1. _____

2. _____

3. _____

New Vocabulary *Use your text to define each term.*

compound _____

_____

chemical bond _____

covalent bond _____

_____

molecule _____

ionic bond _____

metallic bond _____

chemical reaction _____

solution _____

_____

acid _____

base _____

Section 3.2 Combining Matter (continued)

⟨**Main Idea**⟩ ——————— ⟨**Details**⟩ —————————————————

Compounds

Use with pages 66–67.

Organize *information about compounds by completing the graphic organizer below.*

> **Compounds**
> Substances composed of _____ that are
> _____ .

⬇

> **Chemical Bond**
> The forces that _____ are _____
> _____ .

⬇

> **Covalent Bonds**
> Covalent bonds are created when atoms _____ .

⬇

> **Molecule**
> A molecule is _____
> _____ .

⬇

> **Polar Bonds**
> Polar bonds are covalent bonds in which _____
> _____
> _____ .

Ions

Use with page 68.

Organize *information about the formation of ions by completing the flow chart below.*

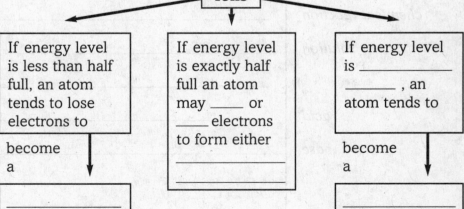

Metallic Bonds

Use with page 68.

Create *a drawing to help you understand metallic bonds. Label positive ions with plus signs and electrons with minus signs.*

Section 3.2 Combining Matter (continued)

Main Idea ———— Details ————————————————

Chemical Reactions

Use with page 70.

Identify *the meaning of symbols in a chemical formula by writing the meaning in words below each symbol.*

2H$_2$	+	O$_2$	→	2H$_2$O

Mixtures and Solutions

Use with page 71.

Summarize *information about mixtures and solutions by completing the graphic organizer below. Answers may vary for examples.*

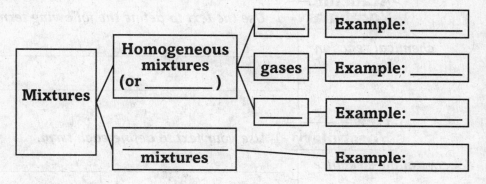

Label *the pH scale to help you understand and remember pH. Write under the acidic and the base parts of the scale two facts about and an example of each.*

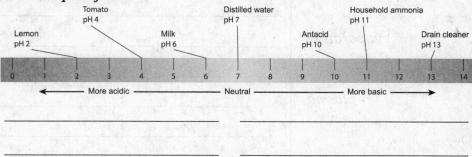

_____   _____

_____   _____

_____   _____

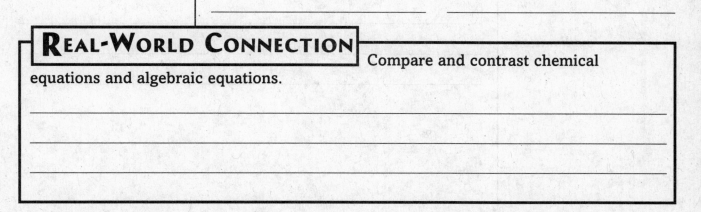

REAL-WORLD CONNECTION

Compare and contrast chemical equations and algebraic equations.

_____

_____

_____

Matter and Change
Section 3.3 States of Matter

Main Idea —— **Details** ——————————

Predict *what you will learn about in Section 3. Read the title and the first paragraph of the section. Then write what you think this section will be about.*

_____

_____

Academic Vocabulary *Use the text to define the following term.*

chemical reaction

_____

_____

New Vocabulary *Use your text to define each term.*

crystalline structure

_____

glass

_____

evaporation

_____

plasma

_____

condensation

_____

sublimation

_____

Section 3.3 States of Matter (continued)

Main Idea ———— **Details** ————————————————

Solids

Use with pages 73–74.

Organize *information about solids by completing the table below.*

Solids	
Definition	**Particles in solids**
• Solids are _____ _____ .	• can be ____ _____ , or _____
• Solids have definite _____ and _____	• are arranged _____ _____
Crystalline Solids	**Glasses**
• Particles are arranged __ _____	• Have no regular internal patterns
• _____ _____ _____	• Are made up of _____ _____ arranged _____
• Most solid substances on Earth are _____ _____	• Form when _____ _____ _____ _____ into a regular pattern.

Liquids

Use with page 74.

Describe *the behavior of particles as a solid becomes a liquid.*

_____

_____

Gases

Use with page 74.

Organize *information about gases by completing the graphic organizer below. Accept all reasonable responses.*

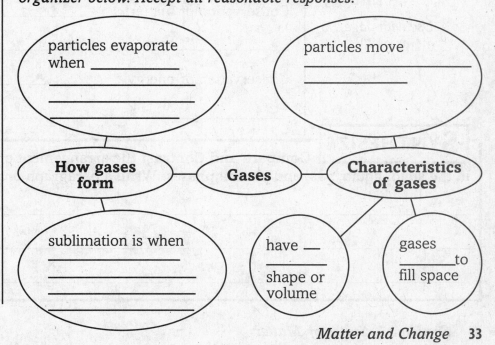

Section 3.3 States of Matter (continued)

⟨**Main Idea**⟩——— ⟨**Details**⟩————————————

Plasma

Use with page 74.

Describe *how plasma forms.*

_____

_____

_____

List *three places where plasma is found.*

1. _____ 2. _____ 3. _____

Changes of State

Use with page 75.

Label *the diagram with the processes that occur when water changes state. Write "sublimation," "evaporation," "condensation," and "freezing." Discuss the transfer of thermal energy for the various state changes in this diagram.*

_____

_____

_____

_____

_____

Solid

Gas   Liquid

Conservation of Matter and Energy

Use with page 75.

Restate *the law of conservation of matter and the law of conservation of energy in your own words.*

Conservation of matter: _____

_____

Conservation of energy: _____

_____

┌─**SYNTHESIZE**─┐

Compare and Contrast the arrangement and behavior of particles in the solid, liquid, gas, and plasma phases. Write a paragraph or make a concept map.

_____

_____

_____

Tie-It-All-Together

Describe *an electron in each of the following situations:*

a. a valence electron on a free atom of an element

b. participating in an ionic bond

c. part of a polar covalent bond

d. conducting electricity through a metal

e. part of a block of ice that melts to liquid water and then evaporates to water vapor

a. _____

_____

_____

_____

b. _____

_____

_____

_____

c. _____

_____

_____

_____

d. _____

_____

_____

_____

e. _____

_____

_____

Matter and Change Chapter Wrap-Up

In the "What I Wanted to Find Out" column, copy the questions you listed in the Chapter Preview. In the "What I Learned" column, write down the answers you discovered as you worked through the chapter.

W What I Wanted to Find Out	L What I Learned
1. _____ _____	1. _____ _____
2. _____ _____	2. _____ _____
3. _____ _____	3. _____ _____

Review *Use this checklist to help you study.*

☐ Study your Science Notebook for this chapter.

☐ Study the definitions of vocabulary words.

☐ Review daily homework assignments.

☐ Reread the chapter and review the tables, graphs, and illustrations.

☐ Review the Section Assessment questions at the end of each section

☐ Look over the Study Guide at the end of the chapter.

SUMMARIZE

After reading this chapter, list three things you have learned about matter.

_____

_____

_____

Minerals
Chapter Preview

Before you read the chapter, use the "What I Know" column to list three things you know about minerals. Then list three questions you have about minerals in the "What I Want to Find Out" column.

K What I Know	W What I Want to Find Out
1. _____ _____	1. _____ _____
2. _____ _____	2. _____ _____
3. _____ _____	3. _____ _____

Science Journal

Think about the Launch Lab you did to observe mineral shapes and record your response in this science journal.

Compare and contrast the shapes of the samples of halite and quartz.

_____

_____

What might account for the differences you observed?

_____

_____

Describe some other physical properties of your mineral samples.

_____

Minerals

Section 4.1 What is a mineral?

⟨Main Idea⟩ ──────────── **⟨Details⟩** ────────────────────────

Scan *Section 1 of your text. Use the checklist below as a guide.*

• Read all headings.

• Read all bold words.

• Read all tables and graphs.

• Look at all the pictures and read their captions.

• Think about what you already know about minerals.

Write one thing you would like to learn about minerals.

_____

⟨Review Vocabulary⟩ *Use your text to define the following term.*

element _____

⟨New Vocabulary⟩ *Use your text to define each term.*

mineral _____

_____

crystal _____

luster _____

hardness _____

cleavage _____

fracture _____

streak _____

specific gravity _____

⟨Academic Vocabulary⟩ *Define the following term.*

restricted _____

Section 4.1 What is a mineral? (continued)

Main Idea	Details

Mineral Characteristics

Use with pages 86–88.

Define *minerals by completing the chart below. Rewrite each part of the definition in your own words.*

A mineral

→ is naturally occurring and inorganic _____ _____ So, ____ is not a mineral.

→ is a solid with a specific composition _____ _____ So, _____ is not a mineral.

→ has a definite crystalline structure _____ _____

Minerals from Magma

Use with page 89.

Describe *how magma forms minerals by completing the flow chart below.*

Magma is _____ _____

When magma is forced upward _____ _____ _____

If the magma reaches Earth's surface before crystallizing, _____ _____ _____

If the magma crystallizes in Earth's heated interior _____ _____ _____

Minerals from Solution

Use with page 89.

Describe *the two ways minerals can form from a solution.*

_____

_____

_____

_____

_____

Minerals

Section 4.2 Types of Minerals

Main Idea ———————— **Details** ———————————————————

Organize *information about mineral identification by completing the outline below.*

Mineral Identification

Use with pages 90–95.

I. _____

    A. Caused by _____ .

    B. Described as _____ or _____ .

II. _____

    A. Defined as _____ .

    B. Measured according to _____ .

III. _____

    A. Determined by _____ .

        1. A mineral has cleavage if _____

        _____ .

        2. A mineral has fracture if _____

        _____ .

IV. _____

    A. Defined as _____

    _____ .

    B. Used only if _____ .

V. Color

    A. Caused by _____ .

    B. Milky appearance can be due to _____

    _____ .

VI. _____

    A. Defined as _____ .

    B. Described as _____ , _____ , _____ , _____ , _____ , or

    _____ .

SYNTHESIZE

Compare and contrast minerals and elements.

_____

_____

_____

Section 4.2 Types of Minerals (continued)

Main Idea ——————— **Details** ————————————————————

VII. _____

    A. Density is a ratio of _____

    _____ .

    B. Density is useful because _____

    _____ .

    C. Specific gravity is ratio of _____

    _____ .

Predict *what you will learn in Section 2. Read the title of the section.*
Write three ways that you might be able to identify a mineral.

1. _____

2. _____

3. _____

Review Vocabulary   *Use your text to define each term.*

chemical bond

_____

_____

New Vocabulary   *Use your text to define each term.*

silicates

_____

_____

tetrahedron

_____

ore

_____

gem

_____

Section 4.2 Types of Minerals (continued)

Main Idea ———— **Details** ——————————————

Draw *three ways that silica tetrahedrons can combine. Circle the diagram that shows how mica, which splits easily into sheets, would bond.*

Mineral Groups

Use with pages 96–98.

Compare *three different mineral groups that oxygen plays a part in.*

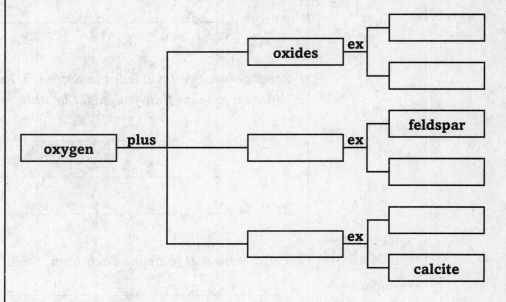

Mineral Uses

Use with page 100.

Create *a concept map to summarize information on ores. Include the definition, uses, processing methods, and examples.*

Section 4.2 **Types of Minerals** (continued)

⟨**Main Idea**⟩ ——

⟨**Details**⟩ ———————————————————————

Gems

Use with page 101.

Identify *why gems are more valuable than other kinds of minerals.*

_____

_____

_____

SYNTHESIZE Describe what factors would have to change for a mineral to be considered an ore.

_____

_____

_____

Minerals Chapter Wrap-Up

In the "What I Wanted to Find Out" column, copy the questions you listed in the Chapter Preview. In the "What I Learned" column, write down the answers you discovered as you worked through the chapter.

W What I Wanted to Find Out	L What I Learned
1. _____ _____	1. _____ _____
2. _____ _____	2. _____ _____
3. _____ _____	3. _____ _____

Review

Use this checklist to help you study.

☐ Study your Science Notebook for this chapter.

☐ Study the definitions of vocabulary words.

☐ Review daily homework assignments.

☐ Reread the chapter and review the tables, graphs, and illustrations.

☐ Review the Section Assessment questions at the end of each section.

☐ Look over the Study Guide at the end of the chapter.

SUMMARIZE After reading this chapter, list three things you have learned about minerals.

_____

_____

_____

Igneous Rocks
Chapter Preview

Before you read the chapter, use the "What I Know" column to list three things you know about igneous rocks. Then list three questions you have about igneous rocks in the "What I Want to Find Out" column.

K What I Know	W What I Want to Find Out
1._____ _____	1._____ _____
2._____ _____	2._____ _____
3._____ _____	3._____ _____

Science Journal

Think about the Launch Lab you did to identify minerals in a sample of rock and record your response in this science journal.

How many different minerals did you observe in the rock?

_____

What minerals can you identify?

_____

_____

Describe the sizes and shapes of the minerals.

_____

Do you see any evidence that these minerals crystallized from molten rock? Explain.

_____

Igneous Rocks

Section 5.1 What are igneous rocks?

Main Idea	Details

Details

Scan *Section 1 of your text. Use the checklist below as a guide.*

- Read all headings.
- Read all bold words.
- Read all tables and graphs.
- Look at all the pictures and read their captions.
- Think about what you already know about igneous rocks.

Hypothesize how igneous rocks are formed. _____

_____

Review Vocabulary   *Use your text to define the following term.*

silicate   _____

_____

New Vocabulary   *Use your text to define each term.*

lava   _____

igneous rock   _____

partial melting   _____

_____

_____

Bowen's reaction series   _____

_____

fractional crystallization   _____

_____

_____

Section 5.1 What are igneous rocks? (continued)

| Main Idea | Details |

Composition and Origins of Magma

Use with pages 112–114.

Organize *information about the composition and formation of magma by completing the outline and the concept web below.*

I. Composition of magma

   A. Magma is _____ .

   B. The elements in magma are _____

      _____ .

      1. The most abundant compound in magma is _____ .

   C. Magmas are classified as _____ , _____ , and _____ .

      1. Their classification depends on _____

      _____ .

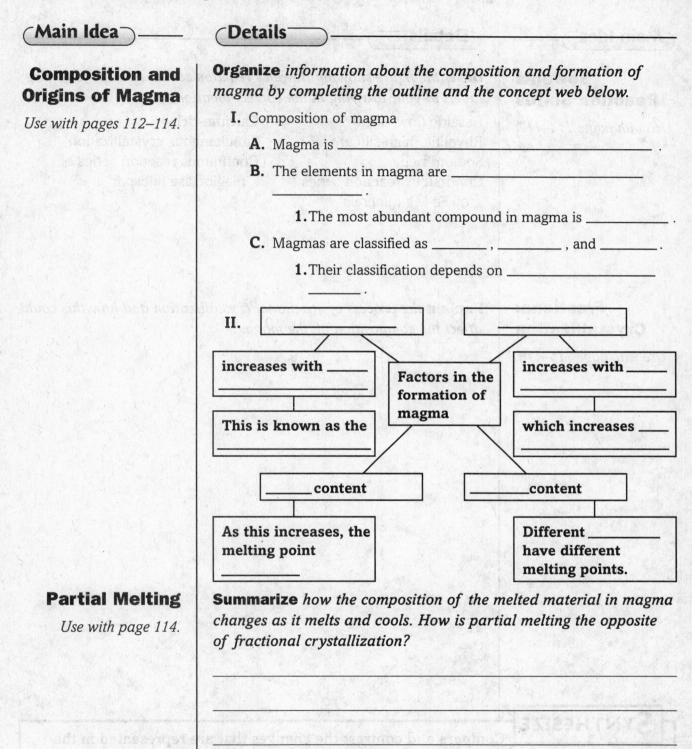

II. _____

increases with _____ _____

This is known as the _____

Factors in the formation of magma

increases with _____ _____

which increases ____

_____ content

_____ content

As this increases, the melting point _____

Different _____ have different melting points.

Partial Melting

Use with page 114.

Summarize *how the composition of the melted material in magma changes as it melts and cools. How is partial melting the opposite of fractional crystallization?*

_____

_____

_____

_____

_____

_____

Section 5.1 What are igneous rocks? (continued)

Main Idea — **Details** —————————

Bowen's Reaction Series

Use with pages 114–117.

Draw *the figure showing Bowen's reaction series. Add the following labels to your drawing in the correct locations.*

Basaltic (low-silica)   Calcium-rich
Rhyolitic (high-silica)   Simultaneous crystallization
Sodium-rich   Continuous reaction series of
Continuous reaction series   plagioclase feldspar
   of Fe-Mg minerals

Fractional Crystallization

Use with page 115–116.

Explain *the process of fractional crystallization and how this could affect the composition of the magma.*

SYNTHESIZE Compare and contrast the changes that are represented in the left and right branches of Bowen's reaction series.

_____

_____

_____

Igneous Rocks
Section 5.2 Classification of Igneous Rocks

Main Idea ——————————— **Details** ————————————————————

Preview *what you will be learning in Section 2. Read the title of the section and the first paragraph on page 118. Describe two ways that geologists use to classify igneous rocks.*

1. _____

2. _____

Review Vocabulary | *Use your text to define the following term.*

fractional crystallization
_____
_____

New Vocabulary | *Use your text to define each term.*

intrusive rock _____

extrusive rock _____

basaltic rock _____

_____

granitic rock _____

texture _____

_____

porphyritic texture _____

_____

vesicular texture _____

pegmatite _____

kimberlite _____

_____

Section 5.2 Classification of Igneous Rocks (continued)

| Main Idea | Details |

Types of Igneous Rocks

Use with pages 118–120.

Summarize *information about intrusive and extrusive igneous rocks by writing each statement in the correct area of the Venn diagram.*

- Coarse-grained
- Form above the surface
- Cool quickly
- Form beneath the surface
- Cool slowly
- Granite is an example
- Fine-grained
- Rhyolite is an example
- Formed by crystallization of magma

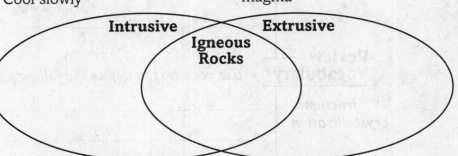

Mineral Composition

Use with pages 118–119.

Interpret *the diagram to list the components of each rock.*

1. basalt: _____

_____

2. diorite: _____

_____

3. pegmatite: _____

_____

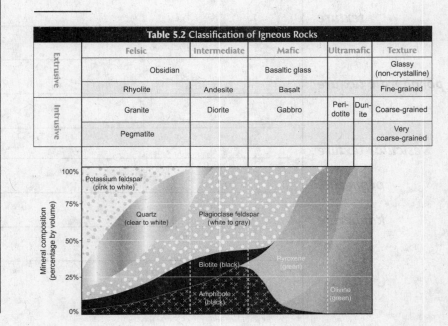

		Felsic	Intermediate	Mafic	Ultramafic		Texture
Extrusive		Obsidian		Basaltic glass			Glassy (non-crystalline)
		Rhyolite	Andesite	Basalt			Fine-grained
Intrusive		Granite	Diorite	Gabbro	Peri-dotite	Dun-ite	Coarse-grained
		Pegmatite					Very coarse-grained

Table 5.2 Classification of Igneous Rocks

Section 5.2 Classification of Igneous Rocks (continued)

⊂Main Idea⊃————— ⊂Details⊃———————————————

Grain Size

Use with pages 119–120.

Describe *the conditions that cause igneous rocks to form large grains or no visible grains.*

_____

_____

_____

_____

_____

Igneous Rocks as Resources

Use with page 123.

List *two characteristics of igneous rocks that make them good building materials.*

1. _____

2. _____

Ore Deposits

Use with page 121.

Organize *the steps in the formation of deposits in veins.*

> The fluid from which magma crystallizes contains high levels of _____
> _____ .

> The fluid fills
> _____
> _____
> ____ .

> The fluid solidifies to form _____ containing metals such as ____,
> ____, ____, _____, and rare elements such as _____ and _____ .

SYNTHESIZE

Identify the characteristic of igneous rocks that you could use to quickly determine whether a rock is extrusive or intrusive. Explain how this characteristic develops.

_____

_____

_____

Igneous Rocks Chapter Wrap-Up

In the "What I Wanted to Find Out" column, copy the questions you listed in the Chapter Preview. In the "What I Learned" column, write down the answers you discovered as you worked through the chapter.

W What I Wanted to Find Out	L What I Learned
1._____ _____	1._____ _____
2._____ _____	2._____ _____
3._____ _____	3._____ _____

Review

Use this checklist to help you study.

☐ Study your Science Notebook for this chapter.

☐ Study the definitions of vocabulary words.

☐ Review daily homework assignments.

☐ Reread the chapter and review the tables, graphs, and illustrations.

☐ Review the Section Assessment questions at the end of each section.

☐ Look over the Study Guide at the end of the chapter.

SUMMARIZE

Hypothesize how N. L. Bowen learned how magma cools in the reaction series. Do you think he performed experiments in a laboratory? Did he make observations in the field? Explain your answer.

_____

_____

_____

Sedimentary and Metamorphic Rocks
Chapter Preview

Before you read the chapter, use the "What I Know" column to list three things you know about sedimentary and metamorphic rocks. Then list three questions you have about them in the "What I Want to Find Out" column.

K What I Know	W What I Want to Find Out
1. _____ _____	1. _____ _____
2. _____ _____	2. _____ _____
3. _____ _____	3. _____ _____

Science Journal

Think of the Launch Lab you did to model sediment layering and record your responses in this Science Journal.

What type of organisms are represented in the sketch?

_____

How did your classmate interpret the drawing?

_____

_____

How did your classmate interpret the number and type of animals that made the tracks?

_____

_____

_____

Sedimentary and Metamorphic Rocks
Section 6.1 Formation of Sedimentary Rocks

Main Idea ─────── **Details** ───────────────────

Scan *Section 1 of your text. Use the checklist below as a guide.*
- Read all headings.
- Read all bold words.
- Read all tables and graphs.
- Look at all the pictures and read their captions.
- Think about what you already know about sedimentary rocks.

Write three questions that come to mind about how sedimentary rocks form.

1. _____

2. _____

3. _____

Review Vocabulary

Use your text to define the following term.

texture

_____

New Vocabulary

In the left margin, write the terms defined below.

_____

pieces of solid material that have been deposited on Earth's surface by wind, water, ice, gravity, or chemical precipitation

_____

physical and chemical processes that transform sediments into sedimentary rocks

_____

when mineral growth cements sediment grains together into rock

_____

horizontal layering in sedimentary rocks

_____

bedding in which the particle sizes become progressively heavier and coarser towards the bottom layers

_____

bedding formed when inclined layers of sediment move forward across a horizontal surface

Section 6.1 Formation of Sedimentary Rocks (continued)

⬭Main Idea⬭ ——————— ⬭Details⬭ —————————————————

Weathering

Use with page 134.

Describe *how weathering produces sediments by completing the concept map below.*

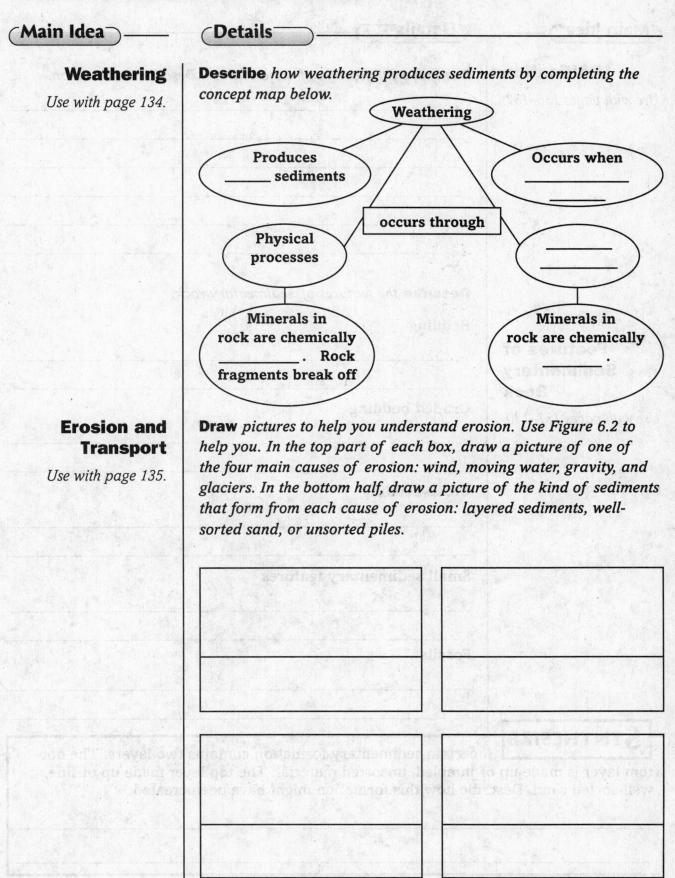

Erosion and Transport

Use with page 135.

Draw *pictures to help you understand erosion. Use Figure 6.2 to help you. In the top part of each box, draw a picture of one of the four main causes of erosion: wind, moving water, gravity, and glaciers. In the bottom half, draw a picture of the kind of sediments that form from each cause of erosion: layered sediments, well-sorted sand, or unsorted piles.*

Section 6.1 Formation of Sedimentary Rocks (continued)

Main Idea | **Details** _____

Lithification

Use with pages 136–137.

Compare *lithification of sand and silt. Describe cementation.*

_____

_____

_____

_____

_____

_____

Describe *the features of sedimentary rock.*

Bedding

_____

Features of Sedimentary Rock

Use with pages 137–140.

_____

Graded bedding

_____

_____

Cross-bedding

_____

_____

Small sedimentary features

_____

_____

Fossils

_____

_____

| **SYNTHESIZE**   A certain sedimentary formation contains two layers. The bottom layer is made up of jumbled, unsorted material. The top layer made up of fine, well-sorted sand. Describe how this formation might have been created.

_____

_____

Sedimentary and Metamorphic Rocks
Section 6.2 Types of Sedimentary Rocks

Main Idea ———— **Details** ————————————

Scan *the objectives for Section 2 in your text. Write three questions that come to mind.*

1. _____

_____

2. _____

_____

3. _____

_____

Review Vocabulary *Use your text to define the following term.*

saturated

_____

_____

New Vocabulary *Use your text to define each term.*

clastic sedimentary rock

_____

_____

clastic

_____

porosity

_____

evaporite

_____

Academic Vocabulary *Define the following term.*

reservoir

_____

_____

Section 6.2 Types of Sedimentary Rocks (continued)

Main Idea ——— **Details** ————————————————

Clastic Sedimentary Rocks

Use with pages 141–143.

Organize *information about the three types of clastic sedimentary rocks.*

Clastic Sedimentary Rocks

→ **Coarse-grained**
Type of fragments:
Types of rock formed:
Formed by:

→ **Medium-grained**
Type of fragments:
Type of rock formed:
Formed by:

→ **Fine-grained**
Type of fragments:
Type of rock formed:

Chemical Sedimentary Rocks

Use with pages 143–144.

Compare *the porosity of medium-grained clastics and fine-grained clastics. Explain the movement of oil and water through each type of rock.*

Medium-grained clastics	Fine-grained clastics

Sequence *the steps in the formation of sedimentary rocks from evaporation. The first one has been completed for you.*

____ Thick layers of evaporites can accumulate as evaporation removes freshwater.

____ Water evaporates and leaves behind dissolved minerals.

__1__ Minerals are dissolved and carried into lakes and oceans by chemical weathering.

____ Layers of chemical sedimentary rocks called evaporites form.

____ The concentration of minerals reaches the saturation point.

____ Crystal grains precipitate out of the solution and settle on the bottom.

Section 6.2 Types of Sedimentary Rocks (continued)

◖Main Idea◗	◖Details◗

Main Idea

Use with pages 143–144.

Details

Summarize *what you read about biochemical sedimentary rocks by completing this table.*

Biochemical Sedimentary Rocks

What forms them	Where they form
How limestone forms	**How coal forms**

Importance of Sedimentary Rocks

Use with page 144.

Organize *information about why sedimentary rocks are important by filling in the graphic organizer.*

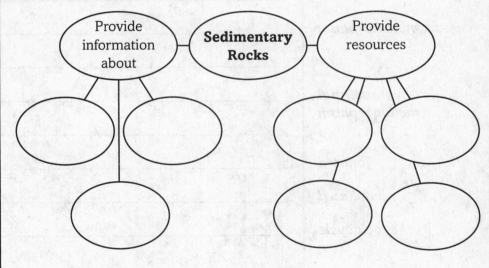

REAL-WORLD CONNECTION A company wants to drill a new oil well. They know that a layer of oil-containing shale is located beneath a thick layer of siltstone. Describe challenges the company might face.

_____

_____

_____

Sedimentary and Metamorphic Rocks
Section 6.3 Metamorphic Rocks

Main Idea ————— **Details** —————————————————

Skim *Section 3 of your text. Read the headings and the illustration captions. Write three questions that come to mind.*

1. _____

2. _____

3. _____

Review Vocabulary   *Use your text to define the following term.*

intrusive   _____

New Vocabulary   *Use your text to define each term.*

regional metamorphism   _____

contact metamorphism   _____

_____

hydrothermal metamorphism   _____

_____

foliated   _____

nonfoliated   _____

rock cycle   _____

_____

Section 6.3 Metamorphic Rocks (continued)

Main Idea — — — — **Details** — — — — — — — — — — — — — — — — — —

Causes of Metamorphism

Use with page 145.

Summarize *information about causes of metamorphism by completing the graphic organizer.*

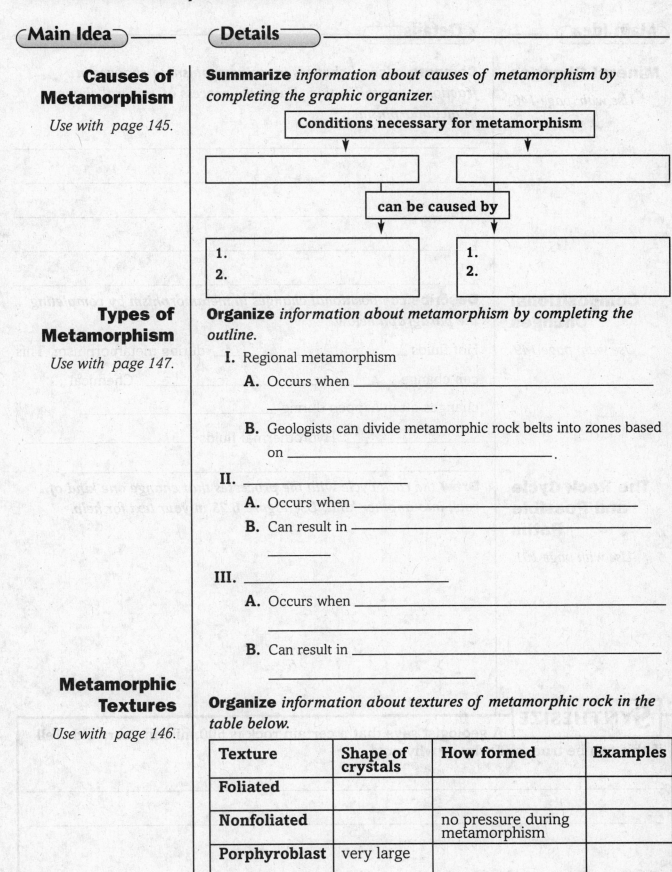

Conditions necessary for metamorphism

can be caused by

| 1. | 1. |
| 2. | 2. |

Types of Metamorphism

Use with page 147.

Organize *information about metamorphism by completing the outline.*

I. Regional metamorphism

   A. Occurs when _____

   _____ .

   B. Geologists can divide metamorphic rock belts into zones based

   on _____ .

II. _____

   A. Occurs when _____

   B. Can result in _____

   _____

III. _____

   A. Occurs when _____

   _____

   B. Can result in _____

   _____

Metamorphic Textures

Use with page 146.

Organize *information about textures of metamorphic rock in the table below.*

Texture	Shape of crystals	How formed	Examples
Foliated			
Nonfoliated		no pressure during metamorphism	
Porphyroblast	very large		

Section 6.3 Metamorphic Rocks (continued)

Main Idea ———— **Details** ——————————————————

Mineral Changes Use with page 146.	**Compare** *mineral changes in metamorphism to changes in fractional crystallization. Name the process of mineral change in metamorphism.* _____ _____ _____ _____ _____
Compositional Changes Use with page 149.	**Describe** *compositional changes in metamorphism by completing the paragraph below.* Hot fluids _____ during metamorphism. This can change _____ . Chemical changes often happen during _____ _____ . Hydrothermal fluids _____ _____ .
The Rock Cycle and Possible Paths Use with page 151.	**Draw** *the rock cycle with the processes that change one kind of rock into another kind. Use Figure 6.23 in your text for help.*

SYNTHESIZE A geologist says that a certain rock is 500 million years old. Tell if this can be true, and why or why not.

_____

_____

Tie-It-All-Together

SYNTHESIZE

Analyze what you have learned in this chapter about how
igneous, sedimentary, and metamorphic rocks are formed. Explain how a rock can
undergo two different processes that change its texture, rock type, or composition.

_____

_____

_____

_____

_____

_____

_____

_____

_____

_____

_____

_____

_____

_____

_____

_____

_____

_____

_____

_____

_____

_____

_____

Sedimentary and Metamorphic Rocks Chapter Wrap-Up

Before you read the chapter, use the "What I Know" column to list three things you know about sedimentary and metamorphic rocks. Then list three questions you have about them in the "What I Wanted to Find Out" column.

W What I Wanted to Find Out	L What I Learned
1. _____ _____	1. _____ _____
2. _____ _____	2. _____ _____
3. _____ _____	3. _____ _____

Review

Use this checklist to help you study.

☐ Study your Science Notebook for this chapter.

☐ Study the definitions of vocabulary words.

☐ Review daily homework assignments.

☐ Reread the chapter and review the tables, graphs, and illustrations.

☐ Review the Section Assessment questions at the end of each section.

☐ Look over the Study Guide at the end of the chapter.

REAL-WORLD CONNECTION

After reading this chapter, list three things you have learned about sedimentary and metamorphic rocks.

_____

_____

_____

Weathering, Erosion, and Soil
Chapter Preview

Before you read the chapter, use the "What I Know" column to list three things you know about weathering, erosion, and soil. Then list three questions you have about them in the "What I Want to Find Out" column.

K What I Know	W What I Want to Find Out
1. _____ _____	1. _____ _____
2. _____ _____	2. _____ _____
3. _____ _____	3. _____ _____

Science Journal

Think about the Launch Lab you did to model interfaces and record your responses in this science journal.

Describe what happened to the sugar cube and the granulated sugar.

_____

_____

Explain why one form of sugar dissolved faster than the other. Infer how you could decrease the time required to dissolve the other form of sugar.

_____

_____

_____

_____

Weathering, Erosion, and Soil

Section 7.1 Weathering

Main Idea	Details
	Scan *the objectives for Section 1 in your text. Write three questions that come to mind.*

1. _____

2. _____

3. _____

Review Vocabulary

Use your text to define the following term. Hypothesize how the term might be important in this chapter.

acid _____

_____

New Vocabulary *Use your text to define each term.*

weathering _____

_____

mechanical weathering _____

frost wedging _____

exfoliation _____

chemical weathering _____

_____

oxidation _____

Academic Vocabulary *Define the following term.*

process _____

_____

Section 7.1 Weathering (continued)

⟨**Main Idea**⟩ ——   ⟨**Details**⟩ ————————————————————

Mechanical Weathering

Use with pages 164–165.

Sequence *the processes by which factors in mechanical weathering break down rocks and minerals. Number the steps in each process 1–4.*

Mechanical Weathering

Temperature

___ Freezing water exerts pressure on the rocks and make them split.

___ Water collects in the cracks of rocks and rock layers.

___ Water thaws and the cycle, called frost wedging, repeats.

___ Water expands as it freezes.

Pressure

___ Successive layers of rock are stripped away in a process called exfoliation.

___ The bedrock surface expands and long cracks form parallel to the surface of the rock.

___ The overlying rock layers are removed and the pressure on the bedrock is reduced.

___ Bedrock at great depths is under pressure from the overlying rock layers.

Chemical Weathering

Use with pages 166–167.

Outline *information about how water, oxygen, and carbon dioxide contribute to chemical weathering to help you understand and remember this cause of erosion.*

  I. Water

    A. Important in chemical weathering because _____

    _____ .

    B. Hydrolysis is _____ .

      1. It occurs in the decomposition _____ .

  II. _____

    A. Like water, it can _____ .

    B. This chemical reaction is called _____ .

  III. _____

    A. Produced by _____ .

    B. Combines with water in the atmosphere to form

    _____ .

    C. Carbonic acid reacts with minerals to _____ .

      1. Limestone caverns can form when _____

      _____ .

Main Idea	Details

Use with page 167.

Organize *information about acid precipitation by completing this graphic organizer.*

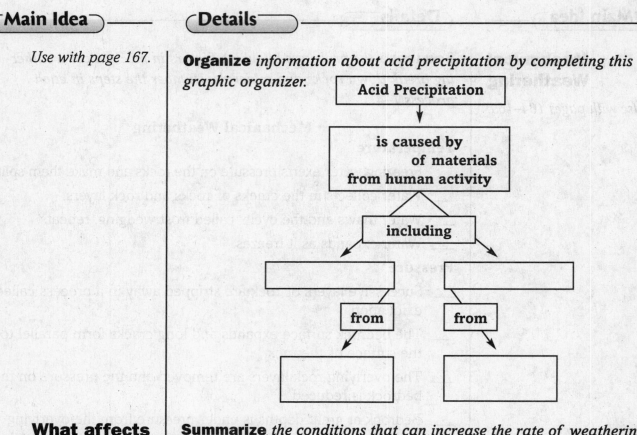

	Acid Precipitation
	↓
	is caused by _____ of materials from human activity
	↓
	including

from _____   from _____

What affects the rate of weathering?

Use with pages 168–170.

Summarize *the conditions that can increase the rate of weathering by completing the table.*

Variables that Affect the Rate of Weathering	
Variable	**Most weathering happens with. . .**
Climate	warm, rainy (chemical) cool, dry (physical)

REAL-WORLD CONNECTION

You have been asked to evaluate a design for a stone monument. The monument will be located in a city that has a warm, rainy climate. The artist has made several errors in judgment. List what they might be, and why. Include information about its placement, design, and the materials you would use to make it.

_____

_____

_____

Weathering, Erosion, and Soil

Section 7.2 Erosion and Deposition

Main Idea ————

Details ————

Consider *the title of Section 7.2 of your book. Write 3 topics that might be discussed in this section.*

1. _____

2. _____

3. _____

New Vocabulary

Use your text to define each term. Make a sketch to help you understand and remember each term.

erosion

_____

_____

_____

deposition

_____

_____

_____

rill erosion

_____

_____

gully erosion

_____

_____

Section 7.2 Erosion and Deposition (continued)

⟨**Main Idea**⟩	⟨**Details**⟩
Gravity's Role in Erosion *Use with page 171.*	**Identify** *gravity's role in erosion. Include examples.* _____ _____ _____
Erosion by Running Water *Use with pages 172–173.*	**Write** *a sentence that tells how each of the following factors affect erosion and/or deposition.* **Streams and rivers** _____ _____ **Deltas** _____ _____ **Ocean currents, waves, and tides** _____ _____ **Dunes and beaches** _____ _____ **Ocean floor and shorelines** _____ _____ **Human development and population growth** _____ _____

Section 7.2 Erosion and Deposition (continued)

| (Main Idea) | (Details) |

Wind Erosion

Use with page 174.

Organize *information about wind erosion by filling in the graphic organizer.*

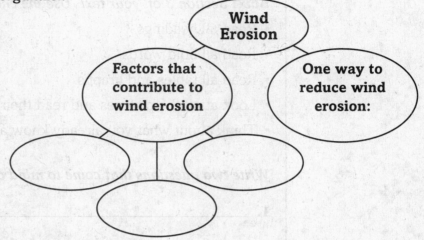

Erosion by Plants, Animals, and Humans

Use with page 175.

Formulate *additional examples of how plants, animals, and humans contribute to erosion. List examples from the book in the left column. Add your own examples in the right column.*

Examples from the book   **Additional examples**

_____   _____

_____   _____

_____   _____

_____   _____

SYNTHESIZE Is erosion a purely destructive force, or does it have constructive aspects as well? Explain your answer and give examples.

_____

_____

_____

_____

Weathering, Erosion, and Soil

Section 7.3 Soil

Main Idea	Details

Scan *Section 3 of your text. Use the checklist below as a guide.*

- Read all headings.
- Read all bold words.
- Read all tables and graphs.
- Look at all the pictures and read their captions.
- Think about what you already know about soil.

Write two questions that come to mind about how soil forms.

1. _____

2. _____

Review Vocabulary *Use your text to define the following term.*

organism _____

_____

New Vocabulary *Use your text to define each term. Make a sketch to help you understand and remember the terms.*

soil _____

_____

residual soil _____

transported soil _____

_____

soil profile _____

soil horizon _____

Section 7.3 Soil (continued)

⟨**Main Idea**⟩ ——   ⟨**Details**⟩ —————————————————————

Development of Soil

Use with page 176.

Sequence *the steps in the development of soil by completing the flow chart.*

┌───┐
│ **Weathering breaks solid bedrock into smaller pieces.** │
└───┘
                            ↓
┌───┐
│   │
└───┘
                            ↓
┌───┐
│   │
└───┘
                            ↓
┌───┐
│   │
└───┘

Soil Composition

Use with page 177.

Write *a paragraph that explains the relationship between a soil and its parent rock.*

_____

_____

_____

_____

_____

_____

Soil Profiles

Use with page 178.

Model *a soil profile in the box below. Use Figure 7.18 and information from your book to help you. If the figure below shows a cross-section of the earth, write each of the labels below in the correct space in the soil profile.*

- High concentrations of organic matter
- Horizon A
- Horizon B
- Sits directly above bedrock
- Subsoils enriched with clay minerals
- Horizon C

┌───┐
│   │
├───┤
│   │
├───┤
│   │
└───┘

Weathering, Erosion, and Soil **73**

Section 7.3 Soil (continued)

Main Idea	Details

Soil Types

Use with pages 179–180.

Examine *Figure 7.21. Describe how various factors result in the soil formed in different areas of the country.*

_____

_____

_____

_____

Soil Fertility

Use with page 182.

Identify *the five factors that affect soil fertility.*

1. _____

2. _____

3. _____

4. _____

5. _____

REAL-WORLD CONNECTION   Every food we eat can be traced back to soil. Take ice cream, for example: ice cream is made with milk; milk comes from cows; cows eat grass; grass grows in soil. So, ice cream comes from soil! Think of a food you like to eat. Then trace the food back to its origin in soil.

_____

Tie-It-All-Together

Weathering and erosion are processes that occur continuously on Earth. Large changes to Earth's surface features often take place slowly, over many years. People do not always consider the effects that these changes can have on the structures they erect. Examine your ideas, thoughts, and opinions about this issue by reading the scenario below and then answering the questions. Support your opinions with logical reasoning.

A large hotel was constructed more than 50 years ago on a high cliff overlooking the Pacific Ocean. Over the years, the cliff has eroded. Ten years ago, a developer built a wharf that reduced the deposition of sand on the beach below the cliff. This increased the rate of erosion of the cliff. Each year, the cliff's edge advances closer and closer toward the hotel. The hotel now sits only 20 m from the edge.

The local government has condemned the hotel, saying that it is no longer safe. The hotel owners have protested the government's decision. They say that the hotel could still be safe for years to come. Furthermore, if the hotel is condemned, they want to be paid for the loss of their property.

Should the hotel owners be allowed to keep the hotel open? Why or why not?

_____

_____

If the hotel did stay open and someone got hurt, would the owners be responsible, or should it just be considered an accident?

_____

_____

If the hotel is closed, should the owners receive any payment? Who would pay them? The local government? The builders of the wharf?

_____

_____

What steps could be taken to ensure that problems like this do not arise in the future?

_____

_____

Weathering, Erosion, and Soil Chapter Wrap-Up

In the "What I Wanted to Find Out" column, copy the questions you listed in the Chapter Preview. In the "What I Learned" column, write down the answers you discovered as you worked through the chapter.

W What I Wanted to Find Out	L What I Learned
1. _____ _____	1. _____ _____
2. _____ _____	2. _____ _____
3. _____	3. _____

Review   *Use this checklist to help you study.*

☐ Study your Science Notebook for this chapter.

☐ Study the definitions of vocabulary words.

☐ Review daily homework assignments.

☐ Reread the chapter and review the tables, graphs, and illustrations.

☐ Review the Section Assessment questions at the end of each section.

☐ Look over the Study Guide at the end of the chapter.

SUMMARIZE   After reading this chapter, describe the connection between weathering, erosion, and soil.

_____

_____

_____

Mass Movements, Wind, and Glaciers
Chapter Preview

Before you read the chapter, use the "What I Know" column to list three things you know about mass movements, wind, and glaciers. Then list three questions you have about mass movements, wind, and glaciers in the "What I Want to Find Out" column.

K What I Know	W What I Want to Find Out
1. _____ _____	1. _____ _____
2. _____ _____	2. _____ _____
3. _____ _____	3. _____ _____

Science Journal

Think about the Launch Lab you did to model sand-slope and record your response in this science journal. Describe how the addition of water affected the sand's ability to be molded in the three samples.

_____

_____

_____

_____

_____

_____

_____

_____

Mass Movements, Wind, and Glaciers

Section 8.1 Mass Movements

Main Idea

Details

Scan *the photos, diagrams, and captions in Section 1 of your text.*
List three topics that are discussed in this section.

1. _____

2. _____

3. _____

New Vocabulary *Use your text to define each term.*

mass movement _____

_____

creep _____

_____

mudflow _____

_____

landslide _____

_____

slump _____

_____

avalanche _____

_____

Section 8.1 Mass Movements (continued)

Main Idea	Details

Factors That Influence Mass Movements

Use with page 195.

List *three variables that influence mass movement.*

1. _____

2. _____

3. _____

Variables That Influence Mass Movements

Use with page 195.

Illustrate *how water can influence soil. Make one drawing to show how water can hold grains together, and one to show how it lubricates between grains.*

Types of Mass Movements

Use with pages 195–199.

Compare *the types of mass movement in the table below. Make a simple sketch of each kind of movement.*

Mass movement	How material behaves	Effects/Results	Sketch
creep			
flow			
slide			
slump			
avalanche			
rock fall			

Section 8.1 Mass Movements (continued)

⟨**Main Idea**⟩ ——————— ⟨**Details**⟩ ——————————————————

Mass Movements Affect People

Use with page 199.

List *three ways human actions increase the destruction caused by mass movements.*

1. _____

2. _____

3. _____

Reducing the Risks

Use with page 200.

Analyze *preventative actions by completing the graphic organizer below.*

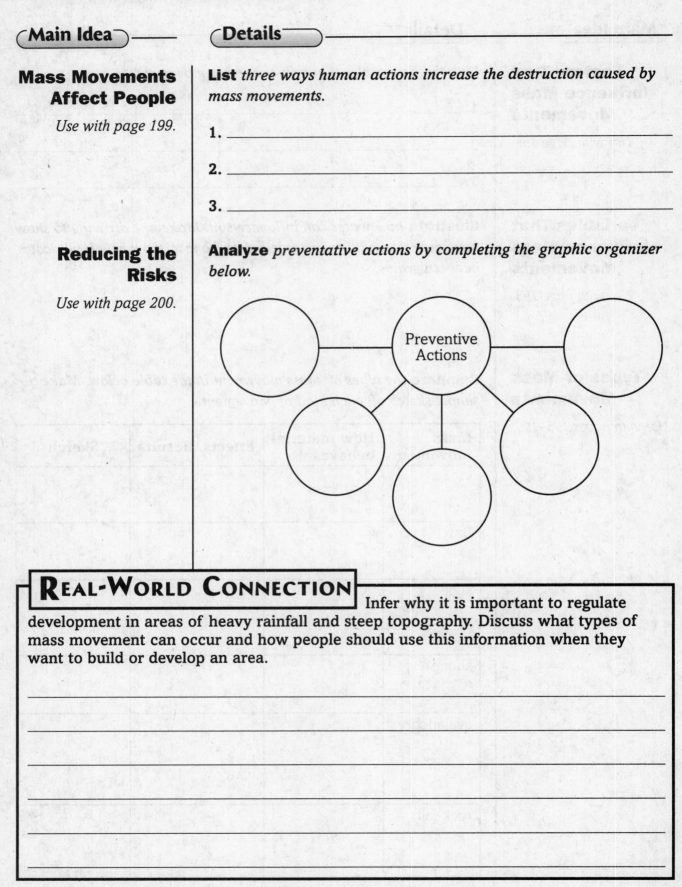

Preventive Actions

┌───┐
| **R**EAL-**W**ORLD **C**ONNECTION   Infer why it is important to regulate |
| development in areas of heavy rainfall and steep topography. Discuss what types of |
| mass movement can occur and how people should use this information when they |
| want to build or develop an area. |
| |
| _____ |
| _____ |
| _____ |
| _____ |
| _____ |
└───┘

Mass Movements, Wind, and Glaciers
Section 8.2 Wind

⟨Main Idea⟩ ——— **⟨Details⟩** ————————————————

Scan *Section 2 of your text. Use the checklist below as a guide.*

• Read all the section titles.

• Read all bold words.

• Look at all figures and photos and read their captions.

• Think about what you already know about rivers and streams.

Review Vocabulary   *Use your text to define the following term.*

velocity   _____

_____

New Vocabulary   *Use your text to define each term.*

deflation   _____

_____

abrasion   _____

_____

ventifact   _____

_____

dune   _____

_____

loess   _____

_____

Academic Vocabulary   *Define the following term.*

migrate   _____

(**Main Idea**) ———— (**Details**) ————————————

Wind Erosion and Transport

Use with page 201.

Organize *information about sediment transport in the Venn diagram to show whether it is a characteristic of saltation, suspension, or both.*

- airborne particles
- areas of low precipitation
- bouncing motion
- moves larger particles
- strong winds
- move smaller particles

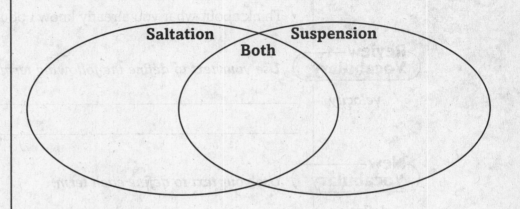

Wind Deposition

Use with page 205.

Draw *diagrams of 4 types of sand dunes in the boxes below. Use Table 8.1 in your text to help. Use an arrow to show the direction the wind is blowing.*

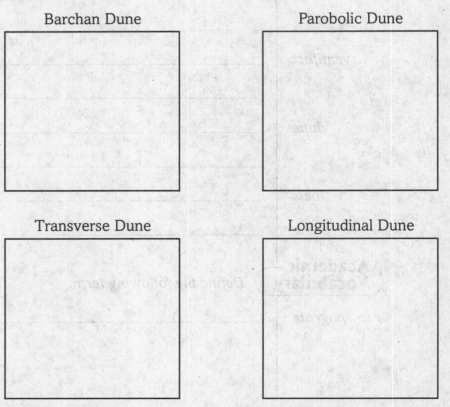

Barchan Dune

Parobolic Dune

Transverse Dune

Longitudinal Dune

Section 8.2 Wind (continued)

◖**Main Idea**◗——— ◖**Details**◗———————————————

Wind Deposition

Use with pages 204–206.

Synthesize *Turn the three bold-faced headings under "wind deposition" into questions. Then use the information from your text to answer your questions.*

Formation of Dunes _____

Question: _____

Answer: _____

_____

Types of Dunes _____

Question: _____

Answer: _____

_____

Loess _____

Question: _____

Answer: _____

_____

┌───┐

REAL-WORLD CONNECTION

Imagine you are a conservation officer for an ocean side state park. How could you explain to the public that they must not damage the dunes? Make up three rules, and give three explanations of what could happen if the rules were not obeyed.

Rule	Consequence

└───┘

Mass Movement, Wind, and Glaciers
Section 8.3 Glaciers

Main Idea   **Details**

Read *the title of the section. Predict three things that might be discussed in this section.*

1. _____

2. _____

3. _____

Review Vocabulary   *Use your text to define each term.*

glacier _____

_____

valley glacier _____

_____

continental glacier _____

_____

cirque _____

moraine _____

outwash plain _____

drumlin _____

_____

esker _____

_____

kame _____

_____

kettle _____

Section 8.3 Glaciers (continued)

⟨**Main Idea**⟩ ————   ⟨**Details**⟩ ——————————————————

Use with page 207.

Glaciers form near _____, and in_____ areas at high elevations.

List *three locations where glaciers can be found today.*

Moving Masses of Ice

Use with page 208.

Compare and Contrast *valley glaciers and continental glaciers in the table below.*

	Valley Glaciers	**Continental Glaciers**
size		
flow begins when . . .		
part of glacier where movement occurs		
locations on Earth		

Glacial Erosion

Use with page 209.

Draw *a diagram or several separate pictures to show the glacial features listed below.*

- waterfall
- cirque
- hanging valley
- horn

Section 8.3 Glaciers (continued)

⟨**Main Idea**⟩ ———   ⟨**Details**⟩———————————————

Glacial Deposition

Use with pages 210–211.

Describe *the differences between glacial till and glacial outwash.*

_____

_____

Describe *the differences between moraines, drumlins, and eskers.*

_____

_____

_____

_____

Sequence *the formation of a kettle lake.*

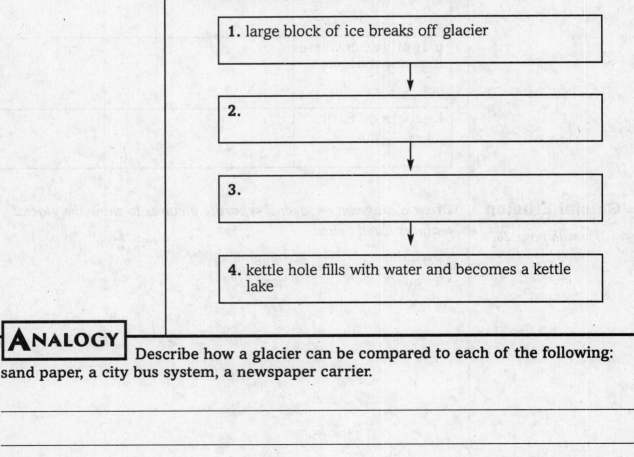

1. large block of ice breaks off glacier

↓

2.

↓

3.

↓

4. kettle hole fills with water and becomes a kettle lake

⟨**ANALOGY**⟩ Describe how a glacier can be compared to each of the following: sand paper, a city bus system, a newspaper carrier.

_____

_____

Tie-It-All-Together

Predict *why mudflows are common when a dry summer with numerous wildfires is followed by an autumn with torrential rains. What precautions can people in this area take to avoid problems?*

_____

_____

_____

_____

_____

Identify *ways that past glacial action in New England and New York State might affect tourism and recreation in those areas.*

_____

_____

_____

_____

Design *a model of one of the types of mass movement mentioned in the chapter. Come up with a plan to stop the erosion or minimize the danger to surrounding areas and homes. Imagine you have an unlimited budget to carry out your plan.*

Mass Movements, Wind, and Glaciers Chapter Wrap-Up

In the "What I Wanted to Find Out" column, copy the questions you listed in the Chapter Preview. In the "What I Learned" column, write down the answers you discovered as you worked through the chapter.

W What I Wanted to Find Out	L What I Learned
1. _____ _____	1. _____ _____
2. _____ _____	2. _____ _____
3. _____ _____	3. _____ _____

Review   Use this checklist to help you study.

☐ Study your Science Notebook for this chapter.

☐ Study the definitions of vocabulary words.

☐ Review daily homework assignments.

☐ Reread the chapter and review the tables, graphs, and illustrations.

☐ Review the Section Assessment questions at the end of each section.

☐ Look over the Study Guide at the end of the chapter.

SUMMARIZE After reading this chapter, list three important ideas about mass movement, wind, and glaciers.

_____

_____

_____

Surface Water
Chapter Preview

Before you read the chapter, use the "What I Know" column to list three things you know about surface water. Then list three questions you have about surface water in the "What I Want to Find Out" column.

K What I Know	W What I Want to Find Out
1. _____ _____	1. _____ _____
2. _____ _____	2. _____ _____
3. _____ _____	3. _____ _____

Science Journal

Think about the Launch Lab you did to model water movement.

Describe what happens to the water after five minutes.

_____

_____

_____

Measure how much water passes through each clump and collects in the plastic shoe box. Explain any differences in the amount of water collected in each plastic shoe box.

_____

_____

_____

Surface Water

Section 9.1 Surface Water Movement

Main Idea — **Details** _____

Scan *Section 1 of your text. Use the checklist below as a guide.*

- Read all the section titles.
- Read all bold words.
- Look at all figures and photos and read their captions.
- Think about what you already know about rivers and streams.

Review Vocabulary   *Use your text to define the following term.*

solution   _____

_____

New Vocabulary   *In the left margin, write the terms defined below.*

_____   water that flows downslope on Earth's surface and may enter a stream, river, or lake; its rate is influenced by the angle of the slope, vegetation, rate of precipitation, and soil composition

_____   land area drained by a stream system

_____   elevated land that divides one watershed or drainage basin from another

_____   when silt and sand are held up and carried along by turbulent stream water

_____   sediments too heavy to remain in solution and are pushed or rolled along the bottom of a stream bed.

_____   a measure of the volume of stream water that flows over a location in an amount of time

_____   potentially devastating natural occurrence in which water spills over the sides of a stream's banks onto adjacent land

_____   broad, flat, fertile area extending from a stream's bank, covered with water during floods

Academic Vocabulary   *Define the following term.*

accommodate   _____

Section 9.1 Surface Water Movement (continued)

| Main Idea | Details |

Runoff

Use with page 225.

Draw *the hydrologic cycle. Include all of the terms listed below. Use Figure 9.1 to help you.*

- condensation
- precipitation
- infiltrates
- groundwater
- transpiration
- ocean
- land
- evaporation
- runoff

Stream Systems Watersheds and Divides

Use with pages 226–227.

Predict *how different variables impact whether precipitation will seep into the ground or become runoff.*

	seep into ground	become runoff
vegetation		
rate of precipitation		
soil composition		
slope		

Sequence *the steps in the development of a stream.*

1. _____

2. _____

3. _____

4. _____

Section 9.1 Surface Water Movement (continued)

Main Idea	Details
Stream Load/Stream Velocity and Carrying Capacity	**Explain** *how minerals, such as calcium carbonate and soluble magnesium compounds found in surface water, compare to the sugar in lemonade.*
Use with pages 228–229.	_____
	_____
	_____
Floodplains/ Floods	**Draw** *a diagram that illustrates the relationship between these terms.*
Use with page 230.	• bedrock
	• floodplain
	• floodplain deposits
	• river channel

REAL-WORLD CONNECTION Imagine you are a developer and have a chance to buy an area for a housing subdivision on a flood plain. Develop a report on this buying option for shareholders, including the risks and benefits of building there.

_____

_____

_____

_____

_____

Surface Water
Section 9.2 Stream Development

Main Idea ————— **Details** ————————————————

Skim *Section 2 of your text. Read the headings and the figure captions. Write three questions that come to mind.*

1. _____

2. _____

3. _____

Review Vocabulary) *Use your text to define the following term.*

abrasion _____

_____

New Vocabulary) *Use your text to define each term.*

stream channel _____

_____

stream bank _____

_____

base level _____

_____

meander _____

_____

_____

delta _____

_____

rejuvenation _____

_____

Section 9.2 Stream Development (continued)

⌐Main Idea⌐	⌐Details⌐

Moving Water Carves A Path

Use with page 232.

Describe *the how moving water can carve a path by completing the paragraph below.*

_____ supplies the water needed for a stream to form. This water first accumulates in an area called the _____ . This is usually a spot high in the _____ . The water accumulates in small _____ at higher elevation and forms _____ streams. Eventually, the water carves out a _____ , which _____ and _____ as more water accumulates. Water is held in a stream by _____ .

Formation of Stream Valleys

Use with page 233.

Sequence *the steps in the formation of stream valleys. The first one has been done for you.*

_____ the stream reaches base level

_____ broader valley with gentle slopes is formed

__1__ a stream erodes steep sided valley

_____ erosion continues along side of V-shaped channel

Meandering Streams

Use with page 234.

Create *a diagram of a meandering stream. Your diagram should include the following:*

• places where deposition occurs

• spots where erosion occurs

• an oxbow lake

• an indication of the area of fastest and slowest flow

Section 9.2 Stream Development (continued)

Main Idea ———— **Details**

Deposition of Sediment

Use with page 236.

Compare and Contrast *an alluvial fan and a delta in the table below.*

	Shape and	Composition	Area and means deposition
Alluvial Fan			
Delta			

Rejuvenation

Use with page 237.

Organize information about the rejuvenation of a stream by completing the cycle diagram below.

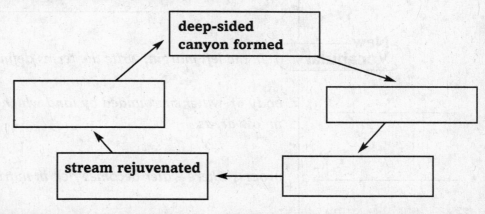

deep-sided canyon formed

stream rejuvenated

SYNTHESIZE

Write the history of a stream. Include information on how and where it was formed, what it experienced during its middle years, and how it aged. Diagrams may be useful in creating this life history story.

_____

_____

_____

_____

_____

_____

_____

Surface Water

Section 9.3 Lakes and Freshwater Wetlands

Main Idea — **Details** ——————————————

Consider the title of Section 3. List three things that might be discussed in this section.

1. _____

2. _____

3. _____

Review Vocabulary Use your text to define the following term.

kettle _____

_____

New Vocabulary In the left margin, write the terms defined below.

_____ body of water surrounded by land which forms in depressions and in low areas

_____ process where water becomes rich in nutrients

_____ land covered in water for most of the year

Academic Vocabulary Define the following term.

undergo _____

_____

Section 9.3 Lakes and Freshwater Wetlands (continued)

<table>
<tr><th>(Main Idea)</th><th>(Details)</th></tr>
</table>

Origins of Lakes

Use with page 238.

List *two kinds of man-made lakes and identify their purpose.*

1. _____

_____

2. _____

_____

Lakes Undergo Change

Use with pages 239–240.

Summarize *the four major ways in which lakes form by describing each process and the type of lake that is formed.*

Streams	Landslides
Glaciers	Groundwater

Analyze *the information on how lakes undergo change. In the question spaces provided, turn the bold-faced headings into questions. Then answer your questions.*

Eutrophication

Question:_____

Answer:_____

_____

Freshwater Wetlands

Question:_____

Answer:_____

Section 9.3 Lakes and Freshwater Wetlands (continued)

Main Idea ————

Lakes Undergo Change

Use with pages 239–241.

Details _____

Identify *two helpful functions that wetlands have.*

1. trap _____, _____, _____ from water sources

2. provide habitats and homes for _____

Create *a concept web to show seven factors, natural and human-contributed, that contribute to the eutrophication of a lake.*

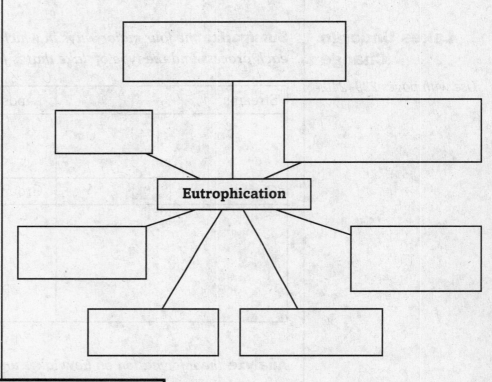

Eutrophication

REAL-WORLD CONNECTION

Create a diagram that explains the probable origin of a lake in your state. Your diagram may include other nearby lakes or landforms to support your hypothesis.

Tie-It-All-Together

Describe *the erosional and depositional abilities of a river.*

_____

_____

_____

_____

Explain *how a floodplain is a dynamic feature.*

_____

_____

_____

_____

Design *a model or several frames (cartoon-strip style) that can show the behavior of a river over its lifetime. Include the different stages and the features that will form and erode over time. Present your model to the class.*

Name _____ Date _____

Surface Water Chapter Wrap-Up

In the "What I Wanted to Find Out" column, copy the questions you listed in the Chapter Preview. In the "What I Learned" column, write down the answers you discovered as you worked through the chapter.

W What I Wanted to Find Out	L What I Learned
1. _____ _____	1. _____ _____
2. _____ _____	2. _____ _____
3. _____ _____	3. _____ _____

Review *Use this checklist to help you study.*

☐ Study your Science Notebook for this chapter.

☐ Study the definitions of vocabulary words.

☐ Review daily homework assignments.

☐ Reread the chapter and review the tables, graphs, and illustrations.

☐ Review the Section Assessment questions at the end of each section.

☐ Look over the Study Guide at the end of the chapter.

SUMMARIZE After reading this chapter, list three things you have learned
about surface water.

_____

_____

_____

Groundwater
Chapter Preview

Before you read the chapter, use the "What I Know" column to list three things you know about groundwater. Then list three questions you have about groundwater in the "What I Want to Find Out" column.

K What I Know	W What I Want to Find Out
1. _____ _____	1. _____ _____
2. _____ _____	2. _____ _____
3. _____ _____	3. _____ _____

Science Journal

Think about the Launch Lab you did to model underground water storage and record your response in this science journal.

Describe how much water is present in the saturated sand.

_____

_____

_____

Calculate the ratio of water volume to the volume of sand. Infer how many liters of water could be stored in a cubic meter of sand.

_____

_____

_____

Groundwater

Section 10.1 Movement and Storage of Groundwater

Main Idea ———————— **Details** ————————————————————

Read the title of Section 1. List three things that might be discussed in this section.

1. _____

2. _____

3. _____

Review Vocabulary Use your text to define the following term.

hydrologic cycle _____

_____

_____

New Vocabulary Use your text to define each term.

infiltration _____

zone of saturation _____

water table _____

zone of aeration _____

_____

permeability _____

_____

aquifer _____

aquiclude _____

spring _____

hot spring _____

geyser _____

Section 10.1 Movement and Storage of Groundwater (continued)

Main Idea ———— **Details** ———————————————

The Hydrosphere

Use with page 252.

Create *a pie graph that compares the percentages of the sources of freshwater. Your graph should include the atmosphere, rivers and streams, groundwater, lakes, and ice caps and glaciers.*

Precipitation and Groundwater

Use with page 253.

Sequence *Number the steps in the movement of groundwater. The first one has been completed for you.*

____ water infiltrates into the ground

_1_ seawater evaporates

____ water becomes part of groundwater

____ precipitation

____ groundwater returns to surface through springs

____ water flows from springs back to the oceans

Groundwater Storage

Use with page 253.

Draw *diagrams of sediments with different porosity.*

Well-Sorted Sediment	Poorly Sorted Sediment

Hypothesize *why poorly sorted sediments might be less porous.*

_____

_____

_____

Section 10.1 Movement and Storage of Groundwater (continued)

Main Idea	Details

The Zone of Saturation

Use with page 254.

Explain *how a kitchen sponge floating in the water compares to the zone of saturation and the water table.*

_____

_____

_____

_____

_____

Groundwater Movement

Use with page 255.

Draw *a diagram similar to Figure 10.4 with an aquifer overlain by an impermeable layer of rock.*

- Show where a reliable well could be dug.

- Draw a well that would be unreliably filled with water.

- Show where a well would be dry.

REAL-WORLD CONNECTION   Draw a fictional topographic map. Show where the water table is at the surface, below the surface, and above the surface of the land. Using graph paper, create a topographic profile (or side view) through a portion of the map. On your profile, indicate the approximate location of the water table.

Groundwater

Section 10.2 Groundwater Erosion and Deposition

Main Idea ———— **Details** ————————————————

Skim through Section 2 of your text. Read the headings and figure captions. Write three questions that come to mind.

1. _____

2. _____

3. _____

Review Vocabulary *Use your text to define the following term.*

hydrolysis _____

New Vocabulary *Use your text to define each term.*

cave _____

_____

sinkhole _____

_____

karst topography _____

_____

stalactite _____

stalagmite _____

_____

Section 10.2 Groundwater Erosion and Deposition (continued)

Main Idea	Details
Dissolution by Groundwater *Use with pages 259–260.*	**Analyze** *cave formation by completing the following sentences.* A _____ is a natural underground opening with a _____ to the surface. Caves form when _____ is dissolved by _____ . The limestone becomes more _____ as it dissolves. Caves usually form in the zone of _____ below the _____ . **Draw** *a diagram to show the landforms found in a karst region. Label the following features on your diagram:* • dry cave • sinkhole • sinking stream • water-filled cave • water table
Groundwater Deposits *Use with pages 261–262.*	**Compare and contrast** *stalagmites and the white "scale" sometimes found in pipes and water kettles.*

Differences	Similarities

Section 10.2 Groundwater Erosion and Deposition (continued)

Main Idea	Details

Groundwater Deposits

Use with pages 261–262.

Skim *the information about groundwater deposits. In the Question spaces provided, turn the bold-faced headings into questions. Then answer the questions.*

Hard Water

Question: _____

Answer: _____

_____

Natural Deposits

Question: _____

_____

Answer: _____

_____

> # REAL-WORLD CONNECTION
>
> *Marble is a metamorphic rock made from limestone that has undergone extreme heat and pressure. Many ancient marble statues are beginning to wear away. Acid rain is the culprit. Explain what is happening to cause this and compare it to the formation of a karst region.*
>
> _____
>
> _____
>
> _____
>
> _____
>
> _____

Groundwater

Section 10.3 Groundwater Wells

Main Idea ————— **Details** —————————————————

Scan *Section 10.3 of your text. Use the checklist below as a guide.*

- Read all the section titles.
- Read all bold words.
- Look at all figures and photos and read their captions.
- Think about what you already know about rivers and streams.

New Vocabulary

In the left margin, write the terms defined below.

_____ *holes dug or drilled to reach reservoir of groundwater*

_____ *difference between original water table level and the level in a pumped well*

_____ *water being added back to zone of saturation*

_____ *well that the water does not need to be pumped from*

Academic Vocabulary *Define the following term.*

transport

_____

_____

Section 10.3 Groundwater Wells (continued)

⟨**Main Idea**⟩—————— ⟨**Details**⟩———————————————————————

Artesian Well

Use with page 264.

Draw *a diagram illustrating how an artesian well forms.*

Wells

Use with pages 263–264.

Describe *how pumping affects a water well by completing the cycle diagram below.*

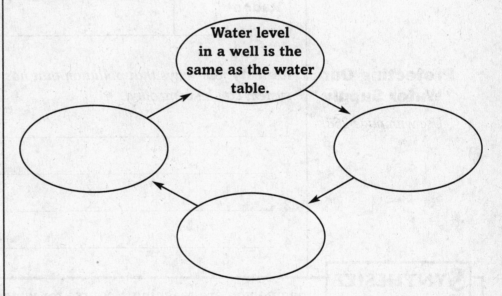

Water level in a well is the same as the water table.

Confined Aquifers

Use with page 264.

Explain concepts related to aquifers by completing the following sentences.

Aquifers that are unconfined are called _____ aquifers.

Confined aquifers are located between two _____ . If a

confined aquifer is under pressure, a well dug in that area may form

an _____ . If water flows naturally from the ground, it may

be an _____ .

Section 10.3 Groundwater Wells (continued)

⟨**Main Idea**⟩———— ⟨**Details**⟩————————————————

Threats to Our Water Supply

Use with pages 265–268.

Summarize *threats to the water supply, causes, and consequences in the table below.*

Threat	Causes	Consequences
Overuse		
Subsidence		
Pollution of Groundwater		
Chemicals		
Salt		
Radon		

Protecting Our Water Supply

Use with page 268.

Identify *four ways that pollution that has entered the groundwater system can be controlled.*

1._____

2._____

3._____

4._____

⟨**SYNTHESIZE**⟩ Imagine you are teaching a course for water-well drillers. List factors that would be important to teach drillers to consider when locating and using water wells, and land features useful for drillers to know about.

_____

_____

_____

_____

Tie-It-All-Together

Describe *some of the problems associated with living in a karst region.*

_____

_____

_____

_____

_____

Explain *how limestone is dissolved by groundwater.*

_____

_____

_____

_____

Design *a model of a cave. Show natural features. Describe animals or plant life. Indicate rock type as well as add the surface of the land to your diagram.*

Groundwater Chapter Wrap-Up

In the "What I Wanted to Find Out" column, copy the questions you listed in the Chapter Preview. In the "What I Learned" column, write down the answers you discovered as you worked through the chapter.

W What I Wanted to Find Out	L What I Learned
1. _____ _____	1. _____ _____
2. _____ _____	2. _____ _____
3. _____ _____	3. _____ _____

Review *Use this checklist to help you study.*

☐ Study your Science Notebook for this chapter.

☐ Study the definitions of vocabulary words.

☐ Review daily homework assignments.

☐ Reread the chapter and review the tables, graphs, and illustrations.

☐ Review the Section Assessment questions at the end of each section.

☐ Look over the Study Guide at the end of the chapter.

SUMMARIZE After reading this chapter, list three things you have learned about groundwater.

_____

_____

_____

Atmosphere
Chapter Preview

Before you read the chapter, use the "What I Know" column to list three things you know about atmosphere Then list three questions you have about atmospheres in the "What I Want to Find Out" column.

K What I Know	W What I Want to Find Out
1. _____ _____	1. _____ _____
2. _____ _____	2. _____ _____
3. _____ _____	3. _____ _____

Science Journal

Think about the Launch Lab you did to model the formation of dew and record your responses in this science journal.

Describe what happened to the outside of the glass in step 3 and step 4.

_____

_____

Relate your observations to the formation of dew.

_____

_____

Graph the temperature of the water during both experiments. Did the results vary with location?

_____

Atmosphere

Section 11.1 Atmospheric Basics

Main Idea ————— **Details** —————————————————————————

Scan *Section 1 of your text. Use the checklist below as a guide.*

• Read all the section titles.

• Read all bold words.

• Look at all figures and photos and read their captions.

• Think about what you already know about atmosphere.

Review Vocabulary) *Use your text to define the following term.*

atmosphere _____

_____

New Vocabulary) *Use your text to define each term.*

troposphere _____

stratosphere _____

radiation _____

conduction _____

convection _____

mesosphere
thermosphere   **Draw** *a sketch showing the relative positions of the mesosphere,*
exosphere   *thermosphere, and exosphere.*

Section 11.1 Atmospheric Basics (continued)

Main Idea ———— **Details** ———————————————————————

Atmospheric Composition

Use with pages 282–283.

Organize *information about the components of the atmosphere using the prompts provided. Use Figure 11.1 and other information from your book to help you.*

The majority of air is made up of _____ (__%) and _____

(__%). _____ and _____ are two important gases

that ____ in amount but are critical in determining the amount of

_____ the atmosphere absorbs. ____ is either absorbed or

released when water _____ .

Three solids in the atmosphere are ____ , ___ , and ____ .

A gas that controls the amount of ultraviolet light that enters the

atmosphere is _____ , a molecule made of three _____ .

Other major gases that can be found in the atmosphere include,

_____ , _____ , _____ , and _____ .

Structure of the Atmosphere

Use with pages 284–286.

Compare and Contrast *the characteristics of the different layers of the atmosphere in the table below.*

Layer	Temperature Variation with Altitude	Composition	Special Characteristics
Troposphere			
Stratosphere			
Mesosphere			
Thermosphere			

Section 11.1 Atmospheric Basics (continued)

⟨**Main Idea**⟩ ──── ⟨**Details**⟩ ──────────────

Solar Fundamentals

Use with pages 286-288.

Summarize *the three methods of energy transfer from the Sun to Earth's atmosphere by completing the following chart.*

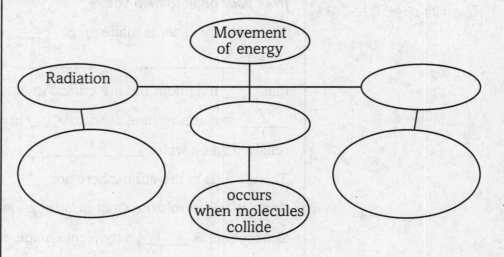

Analyze *how the Sun's energy is affected as it hits Earth's atmosphere. Fill in the table with the five ways it is reflected or absorbed and the percentage of sunlight that goes each way. Use Figure 11.4 to help you.*

Effect					
percent					

┌───┐
│ **REAL-WORLD CONNECTION** │
└───┘
CFCs are banned in the United States because they can cause ozone to degrade. Explain which part of our atmosphere this protects and why that is important.

_____

_____

_____

_____

Atmosphere

Section 11.2 Properties of the Atmosphere

⟨Main Idea⟩	⟨Details⟩
	Read *the title of Section 2. Predict what you think the section might be about.*
	_____
	_____
	_____

⟨Review Vocabulary⟩ *Use your text to define the following term.*

density   _____

⟨New Vocabulary⟩ *In the left margin, write the terms defined below.*

_____   increase in temperature with elevation

_____   amount of water vapor in air

_____   occurs when a volume of air reaches the maximum amount of water vapor

_____   ratio of water vapor in air to how much water vapor that volume of air is capable of holding

_____   temperature to which air must be cooled at constant pressure to reach saturation

_____   extra thermal energy contained in water vapor compared to liquid water

⟨Academic Vocabulary⟩ *Define the following term.*

inversely proportional   _____

_____

Section 11.2 Properties of the Atmosphere (continued)

Main Idea ——— **Details** ————————————————

Measuring Temperature *Use with page 289.*	**Distinguish** *between temperature scales by completing the following sentences.*

There are three temperature scales in use today, the _____ scale is used primarily in the United States. The _____ point of water is 32° on this scale. The SI scale used mainly in science is the _____ scale. _____ on this scale is the lowest temperature any substance can have. The third scale is the _____ scale, sometimes called the centigrade scale because the distance between the freezing and boiling points of water is ____ . Fill in the boiling point of water for each scale; F° ___ , C° ___ , K ___ . Water was chosen as the standard because it is a common substance that exists in all three states of matter at normal temperatures.

Temperature Inversion

Use with page 292.

Draw *a vertical representation of a temperature inversion.*

Air Pressure and Density

Use with page 291.

Predict *whether water would boil on top of a mountain at a higher or lower temperature than at sea level. Explain.*

_____

_____

_____

_____

_____

Section 11.2 Properties of the Atmosphere (continued)

Main Idea — **Details** —

Pressure-Temperature-Density Relationships

Use with page 291.

Sequence *the steps in the formation of temperature inversions.*

1. Cool, clear winter night when the wind is calm

2.

3.

4.

Wind

Use with page 293.

Describe *how wind is created by completing the paragraph below.*

Cool air _____ . This creates an area of _____ .

Warm air _____ . This creates an area of _____ .

Air masses move from areas of _____ to _____ .

This is what creates _____ .

Relative Humidity

Use with page 294.

Draw *a graph and write one sentence to explain why climate zones near the equator are more humid than climate zones closer to the poles. Use the Problem-Solving Lab to help you.*

SYNTHESIZE You are at the beach and notice that in the afternoon there is a cool breeze blowing onto the beach from the ocean. At night, the wind is blowing from the land out to sea. How would you explain to someone why this occurs?

_____

_____

_____

_____

Atmosphere
Section 11.3 Clouds and Precipitation

Main Idea ─────

Details ───────────────────

Scan *the section titles, boldface terms, and diagrams found in this section. Write three topics that will be discussed in this section about moisture in the atmosphere.*

1. _____

2. _____

3. _____

New Vocabulary *Use your text to define each term.*

condensation nuclei _____

_____

orographic lifting _____

_____

cumulus _____

stratus _____

cirrus _____

_____

precipitation _____

coalescence _____

Section 11.3 Clouds and Precipitation (continued)

| Main Idea | Details |

Cloud Formation

Use with pages 297–299.

Describe *clouds by completing the paragraph below.*

_____ can form as warm moist air rises, expands, and cools, and

cooler air _____ in a _____ . Water vapor in the air

condenses on _____ such as dust or ___ . As more

and more droplets collect, a _____ is formed. Clouds also form

from _____ where moist air comes in contact with a

_____ and rises. Another type of cloud forms when _____

of different temperatures collide.

Types of Clouds

Use with page 301.

Analyze *the information on types of clouds. In the question spaces provided, turn the bold-faced headings into questions. Then answer your questions.*

Low Clouds

Question: _____

Answer: _____

_____

Middle Clouds

Question: _____

Answer: _____

_____

High Clouds

Question: _____

Answer: _____

_____

Precipitation

Use with page 302.

Sequence *the formation of precipitation.*

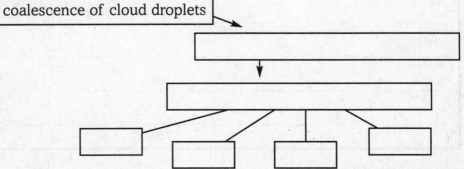

Section 11.3 Clouds and Precipitation (continued)

(**Main Idea**)—————— (**Details**)——————————————————

The Water Cycle

Use with page 303.

Create *your own diagram of the water cycle. Include the following steps in your diagram: evaporation, precipitation, run off, ground-water, and condensation.*

┌───┐
│ **REAL-WORLD CONNECTION**   Clouds are more often formed in areas
of higher air pollution than in areas that have cleaner air. Use what you have learned in
this chapter to describe the process that causes polluted areas to be cloudier.

_____

_____

_____

_____

_____

_____
└───┘

Tie-It-All-Together

Explain why the smog problem is so intense in
Los Angeles. You may want to look at an atlas to answer this question.

_____

_____

_____

_____

_____

Compare the difference between ozone in the atmosphere and ozone near Earth's surface.

_____

_____

_____

_____

Design a model of a greenhouse. Discuss how the heat enters the greenhouse in similar
ways that energy is transferred in the atmosphere.

Atmosphere Chapter Wrap-Up

In the "What I Wanted to Find Out" column, copy the questions you listed in the Chapter Preview. In the "What I Learned" column, write down the answers you discovered as you worked through the chapter.

W What I Wanted to Find Out	L What I Learned
1. _____ _____	1. _____ _____
2. _____ _____	2. _____ _____
3. _____ _____	3. _____ _____

Review *Use this checklist to help you study.*

☐ Study your Science Notebook for this chapter.

☐ Study the definitions of vocabulary words.

☐ Review daily homework assignments.

☐ Reread the chapter and review the tables, graphs, and illustrations.

☐ Review the Section Assessment questions at the end of each section.

☐ Look over the Study Guide at the end of the chapter.

SUMMARIZE

After reading this chapter, list three things you have learned about the atmosphere.

_____

_____

_____

Meteorology
Chapter Preview

Before you read the chapter, use the "What I Know" column to list three things you know about meteorology. Then list three questions you have about meteorology in the "What I Want to Find Out" column

K What I Know	W What I Want to Find Out
1._____ _____	1._____ _____
2._____ _____	2._____ _____
3._____ _____	3._____ _____

Science Journal

Think about the Launch Lab you did to model a cold air mass and record your responses in this science journal. Make a graph showing the temperature changes for each temperature strip.

What happened to the temperature of the air beneath the tray and the air above the tray?

_____

_____

_____

_____

Explain how this model represents a cold air mass.

_____

_____

Meteorology
Section 12.1 The Causes of Weather

Main Idea ————— **Details** ————————————————

Scan *Section 12.1 of your text. Use the checklist below as a guide.*

• Read all the section titles.

• Read all bold words.

• Look at all figures and photos and read their captions.

• Think about what you already know about weather and forecasting.

Write three facts you discovered about causes of weather.

1. _____

2. _____

3. _____

Review Vocabulary *Use your text to define the following term.*

heat _____

New Vocabulary *Use your text to define each term.*

weather _____

climate _____

air mass _____

source region _____

_____

Fill in the flow chart below with key terms from the section.

is long term   which is the current state of the

[ ] → [ ] → [ ]

Section 12.1 The Causes of Weather (continued)

Main Idea ——— **Details** ——————————————————

Weather and Climate

Use with page 314.

Contrast *weather and climate by completing the table below.*

	Weather	Climate
Description		
Short or long term; measure of variations		

A Question of Balance

Use with pages 314–315.

Model *shadows of a person standing in their yard at 6 am, 12 noon, and 6 pm. Assume that north is to the top of the page. Label an approximate degree of latitude your diagram is illustrating.*

Explain *why it is warmest during the middle of the day and cooler in the morning and evening.*

_____

_____

_____

_____

_____

Section 12.1 The Causes of Weather (continued)

⟨**Main Idea**⟩ ———— ⟨**Details**⟩ ——————————————

Air Masses

Use with pages 316–317.

Compare *the major air masses impacting North America.*

Type of Air Mass	Source of Air Mass	Weather Associated with Air Mass
		hot, dry summers
Maritime tropical		
	interior of Canada and Alaska	
		heavy rains in winter on West Coast
Arctic		

What if *you are visiting the deserts of the southwestern United States? A sudden thunderstorm moves in. Hypothesize what type of air mass could have brought that storm, and where it likely originated.*

_____

_____

_____

┌───┐
│ **REAL-WORLD CONNECTION** Determine the type of air mass that
typically plays a role in the weather in your area. Describe the type of weather that commonly is associated with the air mass and where the air mass originates. Describe the seasons in your area which result from that air mass.

_____

_____

_____

_____
└───┘

Meteorology
Section 12.2 Weather Systems

Main Idea _____

Details _____

Scan *Section 2 of your text. Read the headings and the figure captions. Write three questions that come to mind.*

1. _____

2. _____

3. _____

New Vocabulary *Use your text to define each term.*

Coriolis effect _____

_____

trade winds _____

_____

_____

prevailing westerlies _____

_____

_____

polar easterlies _____

_____

jet stream _____

_____

front _____

_____

Academic Vocabulary *Define the following term.*

generate _____

Section 12.2 Weather Systems (continued)

Main Idea	Details

Global Wind Systems

Use with pages 318–320.

Model *the movement of air around the surface of Earth. Use Figure 12.6 to help you. Include the following features:*

- doldrums
- equator
- horse latitudes
- northeast trade winds
- northern hemisphere
- polar easterlies
- prevailing westerlies
- southeast trade winds
- southern hemisphere

Jet Streams

Use with page 321.

Predict *what would happen to the weather in the north-eastern United States if the polar jet stream dipped to the south. How would the weather change if the jet stream moved to the north?*

_____

_____

_____

_____

_____

Section 12.2 Weather Systems (continued)

Main Idea | Details

Fronts

Use with page 322.

Compare *the four main types of fronts.*

Type of front	Map symbol	Description of air movement	Associated Weather
Cold Front			
Warm Front			
Stationary Front			
Occluded Front			

Pressure Systems

Use with page 323.

Differentiate *between high-pressure systems and low-pressure systems. Use the list of characteristics below to complete the Venn diagram.*

- air moves in circular motion
- rising air
- fair weather
- sinking air
- stormy weather
- winds move clockwise in northern hemisphere
- winds move counterclockwise in northern hemisphere

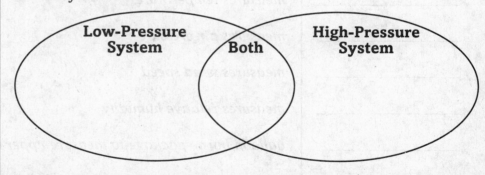

Low-Pressure System Both High-Pressure System

SYNTHESIZE

The doldrums along the equator is an area of very little wind. This gave early sailors a difficult time in their transoceanic trips. How could sailors avoid this trouble spot?

_____

_____

_____

Meteorology
Section 12.3 Gathering Weather Data

Main Idea —— **Details** ——————————————————

Scan *Section 3 of the text. Read the section title, bold words, figures and figure captions. Write four facts you discovered about tropical storms as you scanned the section.*

1. _____

2. _____

3. _____

4. _____

Review Vocabulary

temperature

Use your text to define the following term.

_____

_____

New Vocabulary

Read the definitions below, then write the term for each in the left column.

_____ measures temperature

_____ measures air pressure

_____ measures wind speed

_____ measures relative humidity

_____ balloon borne package to measure upper level atmospheric data

_____ change in wave frequency due to the motion of the wave relative to the observer

Academic Vocabulary

compute

Define the following term.

_____

Section 12.3 Gathering Weather Data (continued)

Main Idea	Details

Surface Data

Use with pages 324–325.

Compare *the different types of instruments that measure surface weather data.*

Instrument	What does it measure?	How does it work?
Barometer		Changes in pressure measured by changes in height of column of mercury
Anemometer		
		Wet- and dry-bulb thermometers
		Data collected by ASOS

Upper Level Data

Use with page 326.

Discuss *collecting weather data in the upper atmosphere.*

Upper-level weather data is collected by a _____ , which is a

series of sensors carried by a _____ . Sensors that measure

_____ , _____ and _____ are carried.

Measurements are sent back by _____ . _____ is very

important when measuring _____ data because it helps

meteorologists determine wind speed and _____ .

Weather Radar

Use with page 327.

Sequence *the way weather radar works in this flow chart.*

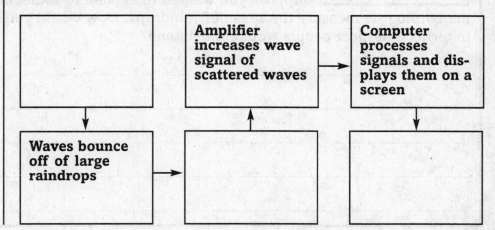

| | → | **Amplifier increases wave signal of scattered waves** | → | **Computer processes signals and displays them on a screen** |

↓ (to "Waves bounce off of large raindrops")

| **Waves bounce off of large raindrops** | → | | | |

Section 12.3 Gathering Weather Data (continued)

⟨**Main Idea**⟩ ——————   ⟨**Details**⟩ ——————————————————

Weather Satellites

Use with page 327.

Organize *information about weather data collection for precipitation tracking by completing the graphic organizer.*

```
┌─────────────────────────┐
│ Scientists can determine│
│ location of precipitation│
│ and clouds using        │
└─────────────────────────┘
         │
    ┌────┴────┐
┌───────┐   ┌───────┐
│       │   │       │
└───┬───┘   └───┬───┘
┌───────┐   ┌───────┐
│       │   │       │
└───────┘   └───────┘
```

Analyze *how infrared imagery is used by meteorologists by completing the following statements.*

Data can be collected in _____.

Infrared imagery detects differences in _____.

Objects show up differently according to the _____.

The temperature of a cloud tells meteorologists about its _____ and _____ .

┌───┐
│ **SYNTHESIZE** Suppose you wanted to explain to someone how   │
│ meteorologists measure the speed of raindrops. How would you explain this procedure │
│ in terms that most people would understand?   │
│   │
│ _____   │
│   │
│ _____   │
│   │
│ _____   │
│   │
│ _____   │
│   │
│ _____   │
└───┘

Meteorology
Section 12.4 Weather Analysis and Prediction

(Main Idea) ——— **(Details)** ————————————————

Consider the title of Section 4. Think of three things that might be discussed in this section.

1. _____

2. _____

3. _____

(Review Vocabulary)

model

Use your text to define the following term. What do you think it has to do with weather?

_____

_____

(New Vocabulary)

Read the definitions below, then write the term for each in the left column.

_____ a record of weather data at a particular site at a particular time

_____ lines of equal pressure

_____ lines of equal temperature

_____ forecast that relies on numerical data

_____ forecast that compares current weather patterns to patterns hat took place in the past

(Academic Vocabulary)

extrapolation

Define the following term.

_____

_____

Section 12.4 Weather Analysis and Prediction (continued)

Main Idea ———— **Details** ————————————————

Surface Analysis

Use with page 329.

Classify *the different types of isopleths found on a weather map.*

```
                    ┌─────────────┐
                    │  isopleths  │
                    └─────────────┘
              ┌───────────┴───────────┐
        ┌───────────┐           ┌───────────┐
        │           │           │           │
        └───────────┘           └───────────┘
         ┌────┴────┐             ┌────┴────┐
    ┌─────────┐ ┌─────────┐ ┌─────────┐ ┌─────────┐
    │ lines of│ │         │ │         │ │indicates│
    │  equal  │ │         │ │         │ │how fast │
    │temperature│ │       │ │         │ │  wind   │
    └─────────┘ └─────────┘ └─────────┘ │ blowing │
                                        └─────────┘
```

Short-Term Forecasts

Use with page 331.

Categorize *the following phrases as either digital or analog forecasting.*

- compare current conditions to past
- looks at conditions in all levels of atmosphere
- monthly or seasonal forecasts
- the more data, the more accurate will be the forecast
- numerical data

Digital	Analog

Section 12.4 Weather Analysis and Prediction (continued)

⟨Main Idea⟩ ——— ⟨Details⟩ ———————————————————

Long-Term Forecasts

Use with page 332.

Summarize *three related factors that cause the accuracy of weather forecasts to decline with time.*

1.

2.

3.

Compare and Contrast *short-term forecasting and long-term forecasting. Discuss reliability, the techniques used, and the type of information that is gathered in each case.*

Short-term forecasting	Long-term forecasting

SYNTHESIZE

Imagine you are reading the newspaper forecast and you want to go downhill skiing the next weekend. Describe the type of information you would like to see on the weather map.

_____

_____

_____

_____

Meteorology Chapter Wrap-Up

In the "What I Wanted to Find Out" column, copy the questions you listed in the Chapter Preview. In the "What I Learned" column, write down the answers you discovered as you worked through the chapter.

W What I Wanted to Find Out	L What I Learned
1. _____ _____	1. _____ _____
2. _____ _____	2. _____ _____
3. _____ _____	3. _____ _____

Review

Use this checklist to help you study.

☐ Study your Science Notebook for this chapter.

☐ Study the definitions of vocabulary words.

☐ Review daily homework assignments.

☐ Reread the chapter and review the tables, graphs, and illustrations.

☐ Review the Section Assessment questions at the end of each section.

☐ Look over the Study Guide at the end of the chapter.

SUMMARIZE After reading this chapter, list three things you have learned about weather systems and weather prediction.

1. _____

2. _____

3. _____

The Nature of Storms
Chapter Preview

Before you read the chapter, use the "What I Know" column to list three things you know about the nature of storms. Then list three questions you have about the nature of storms in the "What I Want to Find Out" column.

K What I Know	W What I Want to Find Out
1. _____	1. _____
_____	_____
2. _____	2. _____
_____	_____
3. _____	3. _____
_____	_____

Science Journal

Think about the Launch Lab you did to replicate thunder and record your response in this science journal.

What happened when the x-side of the balloon was help over the paper? What happened when the opposite side of the balloon was help over the paper?

_____

_____

_____

How does this relate to lightening during an electrical storm?

_____

_____

The Nature of Storms
Section 13.1 Thunderstorms

Main Idea ───── **Details** ──────────────────────────

Read *the objectives on the first page of Section 1. List three questions that come to mind.*

1. _____

2. _____

3. _____

Review Vocabulary *Use your text to define the following term.*

latent heat
_____

_____

New Vocabulary *Use your text to define each term.*

air-mass thunderstorm
_____

_____

mountain thunderstorm
_____

sea-breeze thunderstorm
_____

_____

_____

frontal thunderstorm
_____

_____

stepped leader
_____

_____

return stroke
_____

Section 13.1 Thunderstorms (continued)

Main Idea	Details

How Thunderstorms Form

Use with pages 344–345.

Summarize *three conditions that must be met for a thunderstorm to occur.*

1._____

2._____

3._____

Air-Mass Thunderstorms

Use with page 346.

Draw *diagrams that explain the formation of a mountain thunderstorm and a sea-breeze thunderstorm. Use Figures 13.3 and 11.19 in your text to help you.*

Analyze *Figure 13.1 in your text. Use the diagrams you drew above to explain the reason for the placement of the red and orange zones in Figure 13.1.*

_____

_____

Frontal Thunderstorms

Use with page 346.

Explain *frontal thunderstorms by completing the following sentences.*

The _____ , _____ movement of air in a ____ front can produce a

line of thunderstorms. These thunderstorms can occur at _____ ,

because they do not depend on _____ . Thunderstorms

can also be associated with _____ . These thunderstorms are

usually fairly _____ .

Main Idea	Details

Stages of Development

Use with page 347.

Sequence *the steps in the process of thunderstorm formation. The first one has been completed for you.*

_____ precipitation falls, cooling the air around it

__1__ air rises vertically, creating updrafts

_____ supply of warm, moist air runs out, stopping the updrafts

_____ cloud droplet coalesce, until they are so latent they fall as precipitation

_____ downdrafts form

_____ moisture condenses and releases latent heat

_____ nearly equal amounts of updrafts and downdrafts exist with a cumulonimbus cloud

Compare and contrast *the stages of a thunderstorm. Use Figure 13.4 to help you make detailed sketches of the cumulus stage, the mature stage, and the dissipation stage. Note similarities and differences between the stages.*

_____

_____

REAL-WORLD CONNECTION

Imagine there is a thunderstorm raging outside. You see the lightning flash long before the thunderclap is heard. Describe how you can estimate the distance to the storm.

_____

_____

_____

The Nature of Storms

Section 13.2 Severe Weather

Main Idea ——— **Details** ——————————————————————

Scan *Section 13.2 in your text. Use the checklist below as a guide.*

• Read all the section titles.

• Read all bold words.

• Look at all figures and photos and read their captions.

• Think about what you already know about severe weather and its causes.

Review Vocabulary

Use your text to define the following term.

air mass _____

_____

New Vocabulary

Use your text to define each term.

supercell _____

_____

downburst _____

_____

tornado _____

_____

Fujita tornado intensity scale _____

_____

Academic Vocabulary

Define the following term.

phenomenon _____

_____

Section 13.2 Severe Weather (continued)

Main Idea	**Details**

Severe Thunderstorms

Use with page 350.

Sequence *the steps in the formation of a severe thunderstorm.*

- Strength of storm's updrafts and downdrafts intensifies.
- The cold, high air increases temperature differences.
- A severe storm is formed.
- Cold fronts formed with upper level, low-pressure systems.
- Air becomes more unstable.

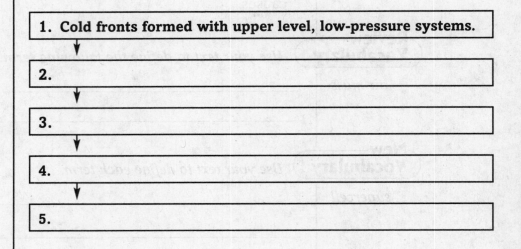

1. Cold fronts formed with upper level, low-pressure systems.

2.

3.

4.

5.

The Fury of the Wind

Use with page 351.

Organize *information about downbursts in the graphic organizer.*

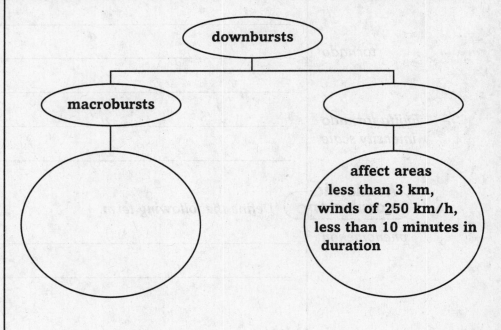

downbursts

macrobursts

affect areas less than 3 km, winds of 250 km/h, less than 10 minutes in duration

Main Idea	Details
Hail, Floods *Use with page 351.*	**Analogy** *An oyster forms a pearl by putting many layers of shell material over a grain of sand or other material. Tell how the formation of a hailstone is similar to the formation of a pearl.* _____ _____ _____ _____
Tornadoes *Use with page 352–353.*	**Define** *Tornado Alley is a real place in the Midwestern United States. This area is more susceptible to tornadoes than other regions of the country. Explain why this is so.* _____ _____ _____ _____ **Model** *tornado formation. Draw a series of pictures to show the formation of a tornado. Use Figure 13.10 as a guide.*

REAL-WORLD CONNECTION

What if you are fishing from a rowboat in the middle of a lake. You notice the clouds building, getting darker, and then it starts to rain. Are you in danger of being struck by lightning? Explain your answer.

_____

_____

The Nature of Storms
Section 13.3 Tropical Storms

Main Idea _____

Details _____

Scan *Section 3 of the text. Read the section title, bold words, figures and figure captions. Write three facts you discovered about tropical storms as you scanned the section.*

1. _____

2. _____

3. _____

Review Vocabulary *Use your text to define the following term.*

Coriolis effect
_____
_____
_____
_____

New Vocabulary *In the left margin, write the terms defined below.*

_____ *large, rotating low-pressure storm that gets its energy from the evaporation of warm ocean water and the release of heat*

_____ *calm center of a hurricane*

_____ *strongest winds of hurricane in a band surrounding the eye*

_____ *classification of hurricanes based on wind speed, air pressure, and potential for property damage*

_____ *mound of ocean water driven toward land by hurricane-force winds*

Academic Vocabulary *Define the following term.*

release _____

Section 13.3 Tropical Storms (continued)

Main Idea	Details

Tropical Cyclones

Use with pages 355–356.

Sequence *the steps in the process of the formation of a hurricane. The first one is done for you.*

____ disturbances produce more precipitation, more energy is released

__1__ as water evaporates from the ocean surface, latent heat is stored

____ the Coriolis effect causes moving air to turn counterclockwise in the northern hemisphere

____ rising air creates a low pressure area at ocean's surface

____ latent heat released, when air begins to rise and clouds form

____ cyclonic rotation of a tropical cyclone

List *the three names used around the world for tropical cyclones, and identify the locations where each is used.*

Name	Locations
A. _____	i. _____
B. _____	i. _____
C. _____	i. _____
	ii. _____
	iii. _____
	iv. _____

Analyze *the formation of tropical cyclones in the outline below.*

I. Two conditions required for a tropical cyclone to form:

A. _____

B. _____

II. Two tropical oceans where tropical cyclones do not form:

A. _____

B. _____

III. Two reasons tropical cyclones do not form in those two places:

A. _____

B. _____

Section 13.3 Tropical Storms (continued)

Main Idea — Details

Model *a hurricane, as seen from above. Use Figure 13.14 in your text as a guide. Your drawing should include:*

- descending air
- direction of rotation
- eye
- eyewall
- warm moist air

Classifying Hurricanes

Use with page 358.

Complete *the following sentences.*

A hurricane usually _____ strength as it moves over ____ or

_____ because it has no access to the _____ from

which it draws its energy. Hurricanes _____ in intensity

over their life cycle as they interact with _____ .

Hurricane Hazards

Use with pages 356–358.

Analyze *why flooding is an additional hazard of hurricanes by completing the statements.*

Flooding occurs due to _____ ,

caused by the _____ .

Floods will be worse if _____ ,

because _____ .

REAL-WORLD CONNECTION

Hurricanes that form during a new moon are potentially more damaging than those that form during the third quarter of the lunar cycle. Explain why this is true.

_____

_____

_____

The Nature of Storms
Section 13.4 Recurrent Weather

(Main Idea) ———— **(Details)** —————————————————————

Consider *the title of the section. Predict three topics that might be discussed in the section.*

1. _____

2. _____

3. _____

(Review Vocabulary) *Use your text to define the following term.*

Fahrenheit scale

_____

_____

(New Vocabulary) *Use your text to define each term.*

drought

_____

_____

_____

heat wave

_____

_____

_____

cold wave

_____

_____

wind-chill index

_____

_____

_____

Section 13.4 Recurrent Weather (continued)

⟨ **Main Idea** ⟩ ——— ⟨ **Details** ⟩ ——————————————

Flood and Droughts

Use with pages 361–362.

Compare and contrast *floods and droughts.*

	Definition	Cause	Impact
Flood			
Drought			

Analyze *the problems associated with large domes of high pressure. Complete the flow chart below.*

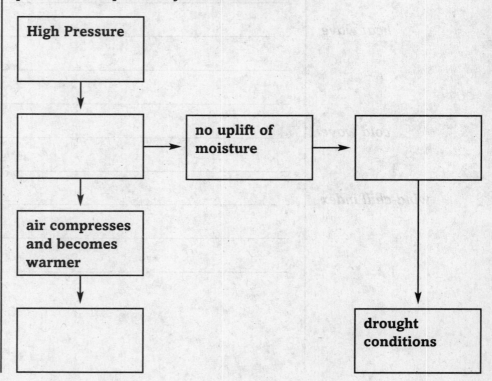

Section 13.4 Recurrent Weather (continued)

(**Main Idea**) ———— (**Details**) ——————————

Cold Waves

Use with pages 362–365.

Analyze *cold waves completing the following flow chart.*

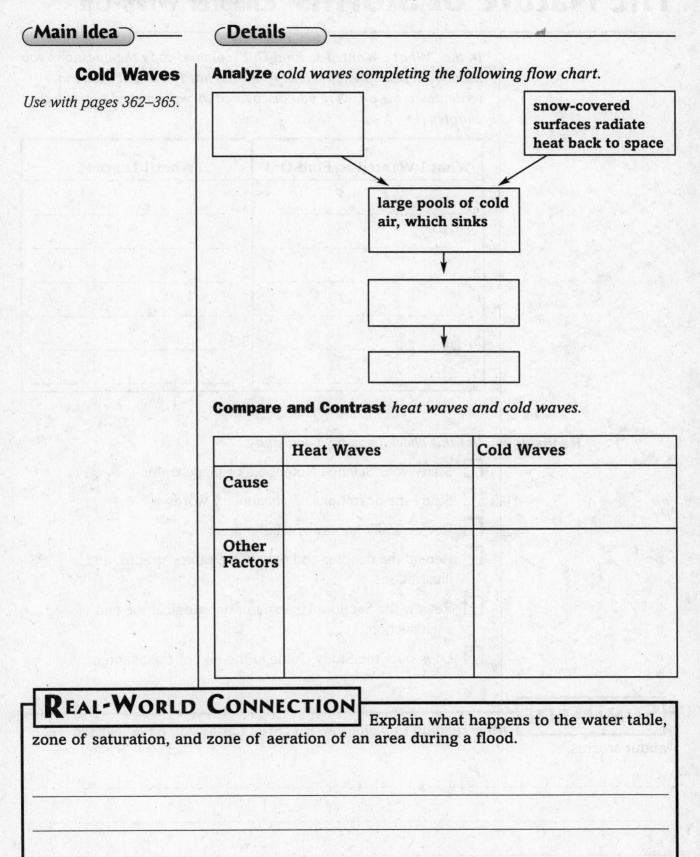

| box (empty) |

snow-covered surfaces radiate heat back to space

large pools of cold air, which sinks

| box (empty) |

| box (empty) |

Compare and Contrast *heat waves and cold waves.*

	Heat Waves	**Cold Waves**
Cause		
Other Factors		

REAL-WORLD CONNECTION Explain what happens to the water table, zone of saturation, and zone of aeration of an area during a flood.

_____

_____

_____

The Nature of Storms Chapter Wrap-Up

In the "What I Wanted to Find Out" column, copy the questions you listed in the Chapter Preview. In the "What I Learned" column, write down the answers you discovered as you worked through the chapter.

W What I Wanted to Find Out	L What I Learned
1. _____ _____	1. _____ _____
2. _____ _____	2. _____ _____
3. _____ _____	3. _____ _____

Review *Use this checklist to help you study.*

☐ Study your Science Notebook for this chapter.

☐ Study the definitions of vocabulary words.

☐ Review daily homework assignments.

☐ Reread the chapter and review the tables, graphs, and illustrations.

☐ Review the Section Assessment questions at the end of each section.

☐ Look over the Study Guide at the end of the chapter.

SUMMARIZE After reading this chapter, list three things you have learned about storms.

_____

_____

_____

Climate
Chapter Preview

Before you read the chapter, use the "What I Know" column to list three things you know about climate. Then list three questions you have about climate in the "What I Want to Find Out" column.

K What I Know	W What I Want to Find Out
1. _____ _____	1. _____ _____
2. _____ _____	2. _____ _____
3. _____ _____	3. _____ _____

Science Journal

Think about the Launch Lab you did to model cloud cover, and record your response to this science journal.

Describe any differences in dew formation that you observed.

_____

_____

How is the umbrella in this activity similar to clouds in the atmosphere?

_____

Based on your observations, infer how temperatures during the night might differ between climates with extensive cloud cover and climates with fewer clouds.

_____

_____

Climate

Section 14.1 Defining Climate

Main Idea	Details

Consider *the title of Section 1. List three things that might be discussed in this section.*

1. _____

2. _____

3. _____

Review Vocabulary

Use your text to define the following term.

jet stream _____

_____

New Vocabulary

Read the definitions below, then write the term on the blank in the left column.

_____ *the study of Earth's climate*

_____ *the standard values for an area*

_____ *the area 23.5° south of the equator and 23.5° north of the equator*

_____ *the area 23.5° and 66.5° north and south of the equator*

_____ *the area located from 66.5° north and south of the equator to the poles*

Academic Vocabulary *Define the following term.*

imply _____

Section 14.1 Defining Climate (continued)

Main Idea	Details

Climate: More than Just Average Weather

Use with page 376.

Explain *climate by completing the graphic organizer below.*

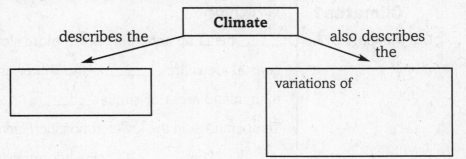

Normals

Use with page 377.

Explain *why most meteorological data are gathered at airports. Then discuss why airports may not be the best location to gather this data.*

_____

_____

_____

What Causes Climates?

Use with page 378.

Sketch *Make a sketch of Earth similar to Figure 14.2 in your text. Label the following in your sketch:*

- two Temperate zones
- Tropics
- Polar zone
- Equator

- 66.5°N and S latitudes
- 23.5°N and S latitudes
- 0° latitude

Section 14.1 Defining Climate (continued)

Main Idea — ### Details

What Causes Climates?

Use with page 378.

Describe *topographic effects by completing the following paragraph.*

_____ heats up and cools down more slowly than ____. Many

coastal areas are _____ in the winter and _____ in the summer

than inland areas of similar _____ .

Temperatures in the lower atmosphere generally decrease with

_____ . Thus, _____ climates are usually cooler than those at

___ ____ . In addition, climates often differ on either side of a moun-

tain. The climate on one side of the mountain—the _____ side—

is usually wet and cool. On the opposite side of the mountain—the

_____ side—the air is ___ , and it warms as it _____ .

Label *the following in the figure below: windward side, leeward
side, moist air, and dry air. Also include arrows to indicate the
direction of the airflow.*

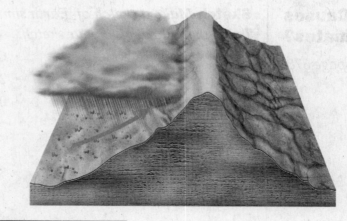

REAL-WORLD CONNECTION Suppose a large reservoir is built in the
middle of an arid region. Infer how normals for that area may change over the long
term as a result.

_____

_____

_____

Climate

Section 14.2 Climate Classification

Main Idea	Details

Scan *Section 2 of your text. Use the checklist below as a guide.*

- Read all section titles.
- Read all bold words.
- Read all graphs and equations.
- Look at all the pictures and read their captions.
- Think about how the word "climate" is used on television weather shows.

Write three facts you discovered about climate as you scanned the section.

1. _____

2. _____

3. _____

Review Vocabulary

Use your text to define the following term.

precipitation _____

New Vocabulary

Use your text to define each term.

Koeppen classification system _____

_____

microclimate _____

_____

heat island _____

_____

Section 14.2 Climate Classifications (continued)

⟨**Main Idea**⟩———— ⟨**Details**⟩————————————————

Use with page 381. **Describe** *the Koeppen classification system by completing the graphic organizer below.*

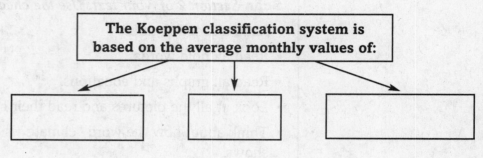

The Koeppen classification system is based on the average monthly values of:

Koeppen Classification System

Use with pages 381–385.

Compare *the five major climates in the Koeppen classification system by completing the table.*

Climate	Characteristics	Examples
Tropical		
Dry		
Mild		
Continental		
Polar		

Section 14.2 Climate Classifications (continued)

Main Idea	Details
Polar Climates Use with page 385.	**Explain** *why precipitation is generally low in polar climates.* _____ _____ _____

Microclimates

Use with page 385.

Describe *the heat-island effect by completing the graphic organizer below.*

The heat-island effect occurs because

This causes mean temperatures in large cities to be

The heat-island effect also causes

This in turn produces

SYNTHESIZE

Scientists tell us that Central Park lowers the temperature of New York City by about 3 degrees. Explain this effect using your knowledge about microclimates.

_____

_____

_____

_____

Climate

Section 14.3 Climatic Changes

Main Idea _____ **Details** _____

Skim *Section 3 of your text. Write three examples of climatic changes that come to mind from reading the headings and illustration captions.*

1. _____

2. _____

3. _____

Review Vocabulary | *Use your text to define the following term.*

glacier _____
_____

New Vocabulary | *Read the definitions below, then write the term on the blank in the left column.*

_____ *periods of time when parts of Earth were covered by glaciers*

_____ *short-term periods of climatic change*

_____ *warm ocean current that develops off the western coast of South America*

_____ *period of very low sunspot activity*

Section 14.3 Climatic Changes (continued)

⟨**Main Idea**⟩ ——— ⟨**Details**⟩ ———————————————————

Ice Ages
Use with page 387.

Analyze *the effects of the glaciers by completing the sentences below.*

_____ is the term that describes extensive glacial coverage. The most recent ice age ended about _____ years ago. During an ice age the average temperature of Earth decreased by ____ . The glaciers eventually retreated, but left evidence in North America. Evidence of the glaciers are the _____ and the _____ Lakes of central New York. We are currently in an _____ period.

Short-Term Climatic Changes
Use with pages 388–389.

Describe *the summer and winter seasons in North America. Also explain the cause for the differences in the two seasons.*

Seasons	Characteristics	Cause
Summer		
Winter		

Sequence *the movements of El Niño, which lead to dramatic climate changes. The first and last steps have been completed for you.*

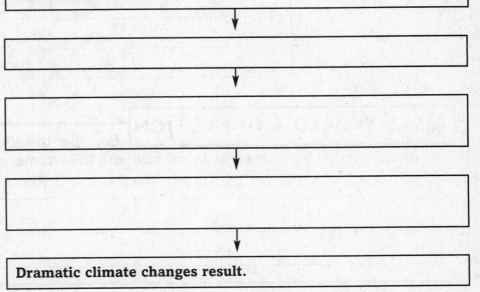

Warm water from the western Pacific moves east toward the South American coast.

↓

↓

↓

↓

Dramatic climate changes result.

Section 14.3 Climatic Changes (continued)

⟨Main Idea⟩——— ⟨Details⟩——————————

Change Can Be Natural

Use with pages 390–392.

Summarize *the effects of increased and decreased solar activity on Earth's climate.*

_____

_____

_____

Sketch *Earth in a circular orbit around the Sun, and then in an elliptical orbit around the Sun. Refer to Figure 14.19 in your textbook for help.*

Explain *how Earth's temperatures are affected when experiencing a circular orbit compared to an elliptical orbit.*

_____

_____

_____

Relate *how major volcanic activity can result in climatic changes.*

_____

_____

_____

REAL-WORLD CONNECTION Consider the effects that El Niño might have on the marine life that come in contact with the warmer ocean currents.

_____

_____

_____

Climate

Section 14.4 Impact of Human Activities

Main Idea ——

Details ——————————————————————

Scan *Section 4 of your text. Use the checklist below as a guide.*

- Read all section titles.
- Read all bold words.
- Read all graphs and equations.
- Look at all the pictures and read their captions.
- Think about what you already know about how humans influence Earth's climate.

Write three facts you discovered about climate as you scanned the section.

1. _____

_____

2. _____

_____

3. _____

_____

Review Vocabulary *Use your text to define the following term.*

radiation _____

_____

New Vocabulary *Use your text to define each term.*

greenhouse effect _____

global warming _____

Section 14.4 Impact of Human Activities (continued)

⟨**Main Idea**⟩———— ⟨**Details**⟩————————————————

The Greenhouse Effect

Use with page 393.

Sketch *the greenhouse effect, similar to Figure 14.23 in your textbook. Use arrows to show how solar radiation reaches Earth and is redirected.*

Global Warming

Use with page 394.

Describe *how an increase in greenhouse gases could affect global temperatures.*

_____

_____

_____

Identify *four possible consequences if the upward trend of world-wide temperatures continues.*

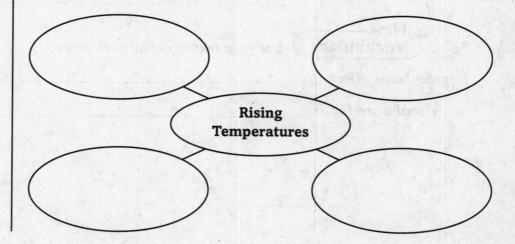

Rising Temperatures

Section 14.4 Impact of Human Activities (continued)

Main Idea　　　　　**Details**

Impact of Human Activities

Use with pages 394–395.

List *two causes for rising levels of atmospheric carbon dioxide.*

1. _____

2. _____

Relate *how deforestation plays a role in increasing levels of atmospheric CO_2.*

_____

_____

_____

Describe *four easy ways to conserve energy.*

1. _____

_____

2. _____

_____

3. _____

_____

4. _____

SYNTHESIZE　In 1988, most of the trees in heavily forested Yellowstone National Park burned. Although it is environmentally sound to let trees burn in order to re-forest, describe the effect that losing so many trees could have on global warming. Explain your reasoning.

_____

_____

Climate Chapter Wrap-Up

In the "What I Wanted to Find Out" column, copy the questions you listed in the Chapter Preview. In the "What I Learned" column, write down the answers you discovered as you worked through the chapter.

W What I Wanted to Find Out	L What I Learned
1._____ _____	1._____ _____
2._____ _____	2._____ _____
3._____ _____	3._____ _____

Review *Use this checklist to help you study.*

☐ Study your Science Notebook for this chapter.

☐ Study the definitions of vocabulary words.

☐ Review daily homework assignments.

☐ Reread the chapter and review the tables, graphs, and illustrations.

☐ Review the Section Assessment questions at the end of each section.

☐ Look over the Study Guide at the end of the chapter.

SUMMARIZE After reading this chapter, list three things you have learned about climate.

_____

_____

_____

Earth's Oceans
Chapter Preview

Before you read the chapter, use the "What I Know" column to list three things you know about physical oceanography. Then list three questions you have about physical oceanography in the "What I Want to Find Out" column.

K What I Know	W What I Want to Find Out
1. _____ _____	1. _____ _____
2. _____ _____	2. _____ _____
3. _____ _____	3. _____ _____

Science Journal

Think about the Launch Lab you did to measure the amount of water on Earth's surface, and record your response to this science journal.

What percentage of the globe's equator is made up of oceans?

_____

What percentage of the globe's equator is made up of land?

_____

Study the globe again. Are the oceans separate bodies of water, or do they interconnect?

_____

_____

_____

Earth's Oceans
Section 15.1 An Overview of Oceans

Main Idea	Details
	Skim *Section 1 of the text. Write three questions that come to mind from reading the headings and the figure captions.*
	1. _____
	2. _____
	3. _____
Review Vocabulary	*Use your text to define the following term.*
lake	_____
	_____
New Vocabulary	*Use your text to define each term.*
oceanography	_____
side-scan sonar	_____
	_____
sea level	_____
Academic Vocabulary	*Define the following term.*
investigate	_____

Section 15.1 An Overview of Oceans (continued)

⟨**Main Idea**⟩————   ⟨**Details**⟩————————————————————

**Modern
Oceanography**

Use with pages 406–407.

Create *a timeline showing the advances in methods used to study Earth's oceans. Your timeline should include specific vessels, technology used, and type of data gathered.*

**Origin of the
Oceans**

Use with pages 408–409.

Analyze *the origins of Earth's oceans.*

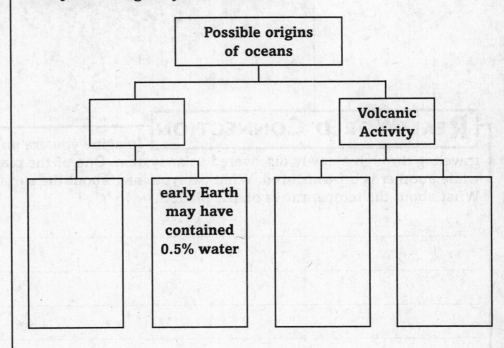

Section 15.1 An Overview of Oceans (continued)

⟨**Main Idea**⟩ ——————— ⟨**Details**⟩ ——————————————

Ocean and Atmospheric Interaction

Use with page 412.

Explain *how the ocean influences weather patterns that form in the atmosphere.*

_____

_____

_____

_____

_____

REAL-WORLD CONNECTION   Imagine you are an astronaut. You are traveling through a newly discovered solar system. One of the planets has a blue tinge while another is brownish-red. What can you infer about the surface of those planets? What about the temperatures on the planet?

_____

_____

_____

_____

Earth's Oceans
Section 15.2 Seawater

Main Idea ——————— **Details** ————————————————————

Scan *Section 2 of your text. Use the checklist below as a guide.*

- Read all the section titles.
- Read all bold words.
- Look at all figures and photos and read their captions.
- Think about what you already know about water in the sea.

Review Vocabulary

Use your text to define the following term. How do you think it relates to seawater?

feldspar _____

_____

_____

New Vocabulary

Use your text to define each term.

salinity _____

estuary _____

temperature profile _____

thermocline _____

_____

Academic Vocabulary

Define the following term.

variation _____

Section 15.2 Seawater (continued)

⊂**Main Idea**⊃ ——————— ⊂**Details**⊃ ——————————————————————

Chemical Properties of Seawater

Use with page 415.

Analyze *how salt is naturally added to and removed from seawater. List two ways salt is added to the sea, and three ways salt is removed from the sea. Use Figure 15.12 and information from your text to help you.*

Salt Added to Sea by. . .	Removed from Sea by. . .

Physical Properties of Seawater

Use with page 417.

Contrast *seawater and fresh water. List three differences.*

1. _____

2. _____

3. _____

Ocean Layering

Use with page 418.

Model *Draw your own graph of variations in ocean water temperatures.*

Be sure to indicate:

• a layer of cold water   • a thermocline   • a layer of warm water

Section 15.2 Seawater (continued)

Main Idea	Details

Water Masses

Use with page 419.

Sequence *the formation of a cold deep water mass.*

____ salt ions accumulate under newly formed ice

____ salty water sinks

__1__ seawater freezes during arctic winter

____ cold, deep water mass migrates toward equator

____ sea ice forms

____ water beneath ice becomes saltier and denser

Organize *information about water masses in the outline below.*

I. Three water masses that account for most of the deep water in the Atlantic Ocean are:

   A. _____

     i. forms when _____

     ii. is the _____ and densest

   B. _____

     i. forms offshore from _____

     ii. overrides the _____

   C. _____

     i. forms when _____

         _____

     ii. least _____ and warmest

     iii. overrides the other ____ masses

II. Two oceans that only contain the two deep antarctic water masses

   A. _____

   B. _____

REAL-WORLD CONNECTION Suppose you go on vacation to Greenland. Your travel companions suggest that the water must be very salty in the ocean surrounding Greenland. How would you respond to this?

_____

_____

Earth's Oceans

Section 15.3 Ocean Movements

Main Idea	**Details**
	Scan *Section 3 of the text. Look at the section titles, bold words, figures, and figure captions. Write three facts you discovered about ocean movements as you scanned the section.*

1. _____

2. _____

3. _____

Review Vocabulary *Use your text to define the following term.*

prevailing westerlies _____

_____

New Vocabulary *Read the definitions below, then write the term for each in the left column.*

_____ *a rhythmic movement that carries energy*

_____ *highest point of a wave*

_____ *lowest point of a wave*

_____ *collapsing waves*

_____ *periodic rise and fall of sea level*

_____ *higher and lower than normal tides caused by solar tides influencing lunar tides*

_____ *when solar tides diminish lunar tides causing high tides to be lower and low tides to be higher*

_____ *caused by wind, affects upper few hundred meters of ocean*

_____ *upward movement of ocean water*

_____ *caused by differences in temperature and salinity of sea water*

Section 15.3 Ocean Movements (continued)

Main Idea	Details

Waves

Use with pages 421–422.

Identify *the characteristics of a wave and show how water moves in waves. Use Figure 15.17 in your text to help you make your own diagrams. Label the crest, half a wavelength, trough, wave base, wave height, wavelength, and where the water movement stops.*

Tides

Use with page 423–425.

Model *the influence the Sun and the Moon have on Earth's tides. Show a spring tide and a neap tide. Use Figure 15.22 to help you. Include the Sun, the Moon in its appropriate phase, Earth, and the oceans in your diagrams.*

Explain *what causes tides by completing the following paragraph.*

Tides are caused by _____ among Earth, the ____ , and the _____ . The fact that gravitational attraction decreases with _____ also plays a role. Unbalanced forces between Earth and the _____ create _____ on opposite sides of Earth. The ____ is farther from Earth, and therefore has a smaller _____ than the _____ despite its greater size.

Name _____ Date _____

⟨**Main Idea**⟩ ──────── ⟨**Details**⟩ ────────────────────

Ocean Currents

Use with page 425.

Compare *the types of ocean currents by completing the table below.*

Ocean Current	Cause	Characteristics	Location
density current		move slowly	
		move fast, follow predictable patterns	

Model *gyres of ocean currents by making a simplified sketch of Figure 15.23. Consider the effects of the Coriolis effect, land masses, and temperature on gyres as you draw.*

Upwelling

Use with page 426.

Sequence *the steps that show how upwelling supports marine ecosystems.*

1. Trade winds move surface water away from the land.

2. _____
   _____

3. _____
   _____

┌───┐
│ **SYNTHESIZE** Imagine Earth was a planet that had two moons. How would this impact the tides on the planet?

_____

_____

_____
└───┘

Tie-It-All-Together

FURTHER INQUIRY

Connect *Ancient mariners traveled the oceans in primitive sailing ships. Imagine you are an ancient sea captain sent out to hunt for food. Explain how knowledge from this chapter would be helpful for navigation and food-finding. Include information on currents, seas and oceans, winds, wave movements, tides, layers, water masses, ice, salinity, and other factors.*

_____

_____

_____

_____

_____

_____

_____

_____

_____

_____

_____

_____

_____

_____

_____

_____

_____

_____

_____

_____

Earth's Oceans Chapter Wrap-Up

In the "What I Wanted to Find Out" column, copy the questions you listed in the Chapter Preview. In the "What I Learned" column, write down the answers you discovered as you worked through the chapter.

W What I Wanted to Find Out	L What I Learned
1. _____ _____	1. _____ _____
2. _____ _____	2. _____ _____
3. _____ _____	3. _____ _____

Review *Use this checklist to help you study.*

☐ Study your Science Notebook for this chapter.

☐ Study the definitions of vocabulary words.

☐ Review daily homework assignments.

☐ Reread the chapter and review the tables, graphs, and illustrations.

☐ Review the Section Assessment questions at the end of each section.

☐ Look over the Study Guide at the end of the chapter.

SUMMARIZE

After reading this chapter, list three things you have learned about oceans. Consider their origin, composition, and the properties of the water.

_____

_____

_____

The Marine Environment
Chapter Preview

Before you read the chapter, use the "What I Know" column to list three things you know about marine environments. Then list three questions you have about them in the "What I Want to Find Out" column.

K What I Know	W What I Want to Find Out
1. _____ _____	1. _____ _____
2. _____ _____	2. _____ _____
3. _____ _____	3. _____ _____

Science Journal

Think about the Launch Lab you did to find out the composition of chalk.

Describe the composition of the powdered chalk.

_____

_____

What is the origin of the chalk?

_____

_____

On what evidence do you base your conclusions?

_____

_____

The Marine Environment

Section 16.1 Shoreline Features

Main Idea ─────

Details ─────

Consider the title of Section 16.1 in your text. Predict three things that might be discussed in this section.

1. _____

2. _____

3. _____

Review Vocabulary

Use your text to define the following term. Hypothesize how it affects the topic of this section.

breaker

_____

New Vocabulary

Use your text to define each term.

beach

_____

_____

wave refraction

_____

_____

longshore bar

_____

longshore current

_____

_____

barrier island

_____

Academic Vocabulary

Define the following term.

estimate

_____

_____

Section 16.1 Shoreline Features (continued)

Main Idea	Details

Erosional Landforms

Use with page 439.

Sequence *the steps in the process of straightening a shoreline. Use Figure 16.3 and information from your text to help you. The first one has been done for you.*

1. waves move faster in deep water than in shallow water

2. _____

3. _____

4. _____

5. _____

6. _____

7. _____

Beaches

Use with page 438.

Describe *how the size and composition of beach sand is determined.*

_____

_____

Predict *what shoreline features would develop if the global sea level were to rise considerably.*

_____

_____

_____

Longshore Currents

Use with pages 440–441.

Analyze *the impact of longshore currents and rip currents on a shoreline by making a diagram similar to Figure 16.6. Your diagram should include longshore bar, shoreline, and an arrow indicating a rip current, an arrow indicating a longshore current, surf zone.*

Section 16.1 Shoreline Features (continued)

⟨**Main Idea**⟩——— ⟨**Details**⟩—————————————————————

Depositional Features of Seashores, Protective Structures

Use with page 442.

Identify *depositional features by drawing a diagram of an eroding shoreline. Identify the following features on your diagram. Use Figure 16.8 and information from your text to help you.*

- bay
- mainland beach
- tombolo
- baymouth bar
- spit
- lagoon
- barrier island

Changes in Sea Level

Use with page 445.

Analyze *changes in sea level by completing the graphic organizer below.*

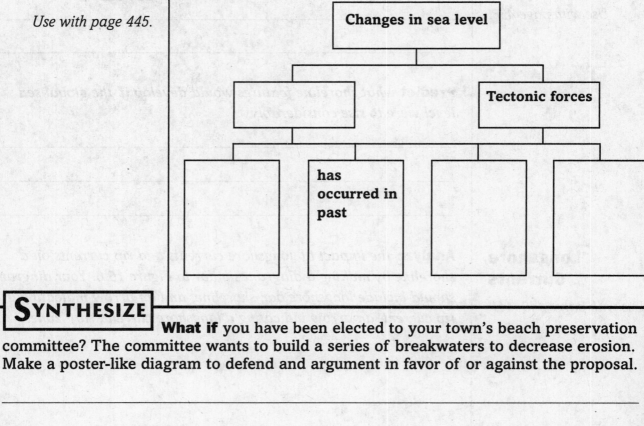

| Changes in sea level |

| | Tectonic forces |

| has occurred in past |

⟨**SYNTHESIZE**⟩ **What if** you have been elected to your town's beach preservation committee? The committee wants to build a series of breakwaters to decrease erosion. Make a poster-like diagram to defend and argument in favor of or against the proposal.

_____

_____

The Marine Environment

Section 16.2 The Seafloor

Main Idea ————

Details —————————————————————————

Scan *Section 2 in your text. Look at the headings, photos, illustrations, and captions. Write three facts you discovered about the seafloor.*

1. _____

2. _____

3. _____

New Vocabulary

Read the definitions below, then write the term for each in the left column.

_____ *submerged parts of continents*

_____ *shallowest part of a continental margin extending seaward*

_____ *area where seafloor drops quickly to depths of several kilometers*

_____ *rapidly flowing water currents carrying heavy loads of sediment*

_____ *accumulation of deposits from turbidity currents*

_____ *plains with thick deposits of marine sediment above basaltic rock*

_____ *deepest part of ocean basin*

_____ *most prominent feature in ocean basin*

_____ *submerged basaltic volcanoes*

_____ *large, extinct, basaltic volcanoes with flat, submerged tops*

Section 16.2 The Seafloor (continued)

Main Idea	Details

Oceanic and Continental Crust

Use with page 447.

Compare *oceanic and continental crust by completing the table below.*

Crust	Average thickness	Location
Continental	40 km	
Oceanic		

Continental Shelves, Continental Slopes, Ocean Basins

Use with pages 447–448.

Identify *the features in the list below on a diagram of the continental margin. Use Figure 16.16 to help you.*

- continental rise
- continental slope
- continental shelf
- shelf break
- abyssal plain
- submarine canyons
- continental margin

Section 16.2 The Seafloor (continued)

⟨**Main Idea**⟩ ———— ⟨**Details**⟩ ——————————————

Seafloor Volcanoes

Use with page 452.

Compare *seamounts and hydrothermal vents in the Venn diagram below. Use the following phrases to fill in the circles.*

• hole in seafloor where magma erupts
• formed on ocean floor
• bottom of rifts in mid-ocean ridges
• in area of no current volcanism
• extinct volcanoes

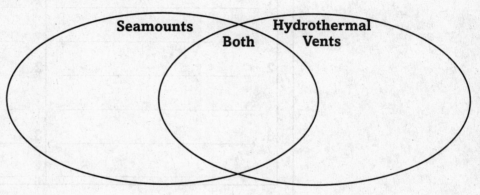

Seamounts   Both   Hydrothermal Vents

Marine Sediments

Use with pages 453–454.

Identify *the three types of marine sediment. Describe each of them.*

1. _____

2. _____

   _____

   _____

3. _____

   _____

REAL-WORLD CONNECTION Hypothesize where you would you find the youngest rock on the planet. Support your reasoning with information from the section.

_____

_____

The Marine Environment Chapter Wrap-Up

In the "What I Wanted to Find Out" column, copy the questions you listed in the Chapter Preview. In the "What I Learned" column, write down the answers you discovered as you worked through the chapter.

W What I Wanted to Find Out	L What I Learned
1. _____ _____	1. _____ _____
2. _____ _____	2. _____ _____
3. _____ _____	3. _____ _____

Review *Use this checklist to help you study.*

☐ Study your Science Notebook for this chapter.

☐ Study the definitions of vocabulary words.

☐ Review daily homework assignments.

☐ Reread the chapter and review the tables, graphs, and illustrations.

☐ Review the Section Assessment questions at the end of each section.

☐ Look over the Study Guide at the end of the chapter.

SUMMARIZE After reading this chapter, list three things you have learned about the marine environment, including shoreline features and the seafloor.

_____

_____

_____

Plate Tectonics
Chapter Preview

Before you read the chapter, use the "What I Know" column to list three things you know about plate tectonics. Then list three questions you have about plate tectonics in the "What I Want to Find Out" column.

K What I Know	W What I Want to Find Out
1. _____ _____	1. _____ _____
2. _____ _____	2. _____ _____
3. _____ _____	3. _____ _____

Science Journal

Think about the Launch Lab you did to determine fault movement in California, and record your response to this science journal.

Infer what might be causing these large pieces of land to move.

_____

_____

_____

_____

_____

_____

_____

Plate Tectonics
Section 17.1 Drifting Continents

Main Idea ———————— **Details** ————————————————————

Scan *Section 1 of your text. Write three questions that come to mind from reading the headings and figure captions.*

1. _____

2. _____

3. _____

Review Vocabulary *Use your text to define the following term.*

hypothesis _____

New Vocabulary *Use your text to define each term.*

continental drift _____

_____

Pangaea _____

_____

Name _____ Date _____

Section 17.1 Drifting Continents (continued)

Main Idea	Details

Early Observations

Use with pages 468–469.

Analyze *the growth of the continental drift theory by summarizing contributions by each person in the table below.*

Name			Alfred Wegener
Nationality		Austrian	
Date	late 1500s		
Profession			
Observation		built on matching coastline observations by other people	built upon observations by other people
Proposed Explanation			

Continental Drift

Use with pages 469–470.

Create *three diagrams to show how it is thought that the continents drifted. Use Figure 17.2 to help you draw what Earth looked like 200 million years ago, 135 million years ago, and in the present.*

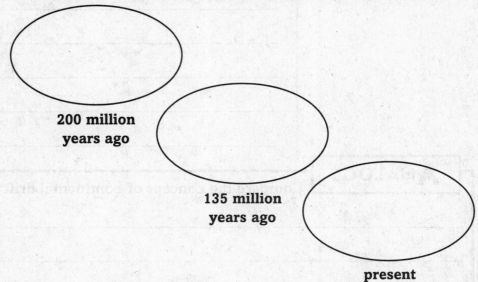

200 million years ago

135 million years ago

present

Section 17.1 Drifting Continents (continued)

Main Idea ———— **Details** —————————————————

Analyze *the evidence Wegener had supporting his hypothesis of continental drift.*

Evidence	Concept/Idea	Examples of evidence
Rock evidence		
Fossil evidence		
Climate evidence		

A Rejected Hypothesis

Use with page 472.

Identify *two reasons why scientists rejected Wegener's hypothesis of continental drift.*

_____

_____

_____

_____

_____

_____

_____

ANALOGY

Compare the concept of continental drift to a jigsaw puzzle.

_____

_____

Plate Tectonics
Section 17.2 Seafloor Spreading

Main Idea _____

Details _____

Consider the objectives of Section 2. Write three questions that come to mind from reading these statements.

1. _____

2. _____

3. _____

Review Vocabulary · Use a dictionary to define the following term.

basalt _____

_____

New Vocabulary · Use your text to define each term.

magnetometer _____

_____

magnetic reversal _____

_____

paleomagnetism _____

_____

isochron _____

seafloor spreading _____

Section 17.2 Seafloor Spreading (continued)

⟨**Main Idea**⟩———— ⟨**Details**⟩————————————————

Magnetism

Use with page 476.

Explain *why the seafloor exhibits magnetic reversal, but the continents do not.*

_____

_____

_____

_____

_____

Ocean Floor Topography

Use with pages 474–475.

Organize *information about ocean floor topography by completing the following flow chart.*

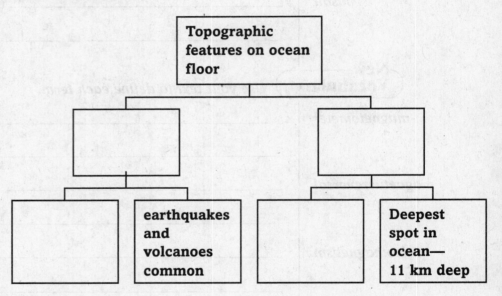

Topographic features on ocean floor

earthquakes and volcanoes common

Deepest spot in ocean— 11 km deep

Ocean Rocks and Sediments

Use with page 475.

Predict *where the oldest rocks in the Atlantic Ocean are. Predict where the youngest rocks are. Use maps in your text to help you answer this question.*

_____

_____

_____

Section 17.2 Seafloor Spreading (continued)

Main Idea ——— Details ————————————————————

Magnetism

Use with page 476.

Illustrate *magnetic symmetry around an ocean ridge. Use Figures 17.12 and 17.13 to help you. Show the following features:*
- ocean ridge
- areas of reversed polarity
- areas of normal polarity
- relative ages of rocks

Describe *how a scientist would know how to draw isochrons to connect basalt flows on land with the ocean floor.*

_____

_____

_____

Seafloor Spreading

Use with page 479.

Organize *information about seafloor spreading by completing the following cycle diagram.*

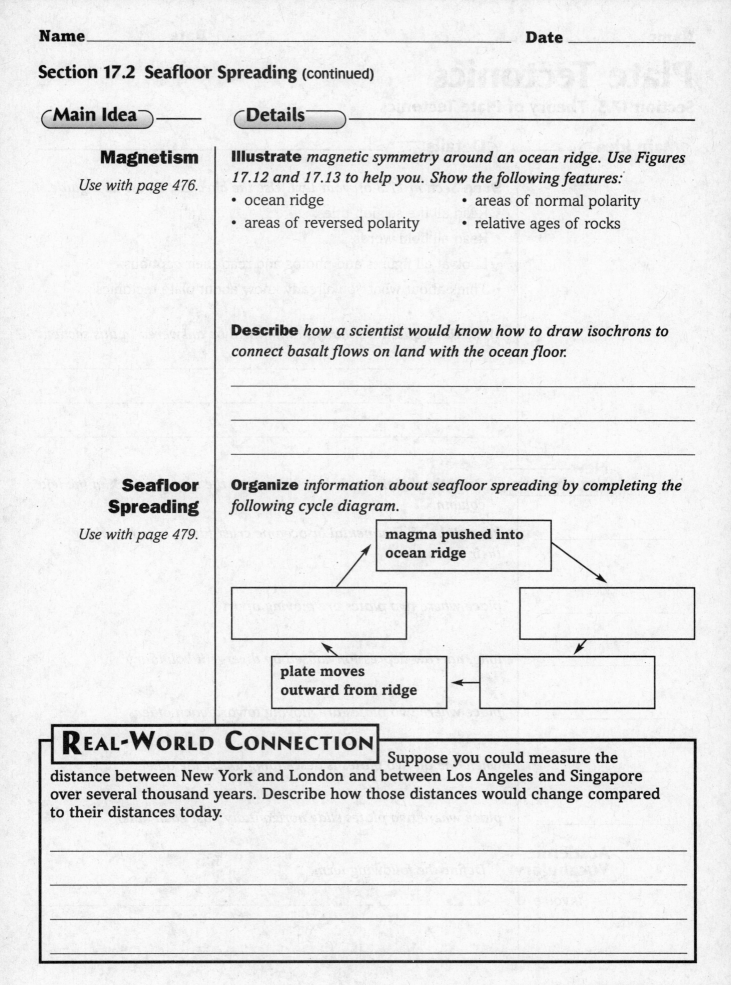

magma pushed into ocean ridge

plate moves outward from ridge

REAL-WORLD CONNECTION Suppose you could measure the distance between New York and London and between Los Angeles and Singapore over several thousand years. Describe how those distances would change compared to their distances today.

_____

_____

_____

_____

Plate Tectonics
Section 17.3 Theory of Plate Tectonics

(Main Idea) ——— **(Details)** ————————————————————

Scan *Section 17.3 of your text. Use the checklist below as a guide.*

- Read all the section titles.
- Read all bold words.
- Look at all figures and photos and read their captions.
- Think about what you already know about plate tectonics.

Write three questions that you think will be answered in this section.

_____

_____

_____

(New Vocabulary) *Read the definitions below, then write the term for each in the left column.*

_____ *huge slabs of continental or oceanic crust fitting together along their edges*

_____ *place where two plates are moving apart*

_____ *long, narrow depression caused by divergent boundary*

_____ *place where two plates are moving toward each other*

_____ *when one of two plates is descending under the other*

_____ *place where two plates slide horizontally past each other*

(Academic Vocabulary) *Define the following term.*

involve _____

Section 17.3 Theory of Plate Tectonics (continued)

⟨ **Main Idea** ⟩——— ⟨ **Details** ⟩————————————————

Plate Boundaries

Use with pages 481–489.

Illustrate *a divergent plate boundary and the three types of convergent plate boundaries. Include the crust, asthenosphere, lithosphere, upper mantle, and ocean ridge on the divergent plate boundaries, and trench, volcanoes, upper mantle, oceanic crust, and continental crust on the convergent plate boundary diagrams.*

Below your diagrams, list at least four products or results at each type of plate boundary. Use Figures 17.17 and Table 17.1 to help you.

Divergent Plate Boundary

Oceanic-Oceanic Convergent

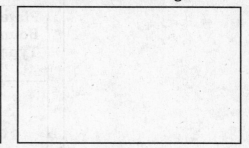

Oceanic-Continental Convergent

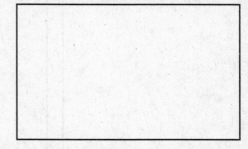

Continental-Continental Convergent

products/results:

_____

_____

_____

_____

products/results:

_____

_____

_____

_____

Section 17.3 Theory of Plate Tectonics (continued)

⟨**Main Idea**⟩ ——— ⟨**Details**⟩ ————————————————

Plate Boundaries

Use with pages 481–485.

Predict *Imagine that a fence perpendicularly crosses a transform plate. Draw diagrams of the fence and the two plates before and after the plates move. Use arrows to show the direction of movement of the plates.*

Compare *the three types of plate boundaries and their characteristics in the table*

Plate Boundary Type	Location	Impact on plates features involved	Associated geologic phenomena	New features formed?
	most on ocean floors	two oceanic plates moving apart		
Convergent		oceanic crust sinks beneath oceanic crust; oceanic crust sinks beneath continental crust; continental crust and continental crust collide		
Transform	two plates moving past each other, plate edges			n/a

SYNTHESIZE The oldest rocks on the continents are almost four billion years old. The oldest rocks on the ocean floor are less than 200 million years old. Analyze the reasons for this difference.

_____

_____

Plate Tectonics

Section 17.4 Causes of Plate Motions

◖Main Idea◗ ———————— **◖Details◗** ———————————————————

Consider *the title of Section 4. List three things that might be discussed in this section.*

1. _____

2. _____

3. _____

◖Review Vocabulary◗ *Use your text to define the following term.*

convection _____

_____

◖New Vocabulary◗ *Use your text to define each term.*

ridge push _____

_____

slab pull _____

_____

◖Academic Vocabulary◗ *Define the following term.*

mechanism _____

_____

Section 17.4 Causes of Plate Motions (continued)

Main Idea	Details

Mantle Convection

Use with page 488.

Identify *two factors that cause convection. Describe the movement of a convection current and sketch a current.*

_____

_____

_____

_____

_____

Use with pages 487–488.

Compare *the processes of ridge push and slab pull by filling in the table below.*

	Type of plate boundary	Description of movement	Impact on plate tectonics
Ridge push			
Slab pull			

Explain *mantle convection currents by completing the following sentences.*

_____ currents in the _____ are thought to be the mechanism behind _____ . The _____ of the mantle is like soft _____ , so it can be moved like a fluid by convection currents.

_____ energy inside Earth's _____ starts the convection currents moving. Hot mantle material is _____ than the cooler mantle material. Therefore, it _____ . Cooler parts of the mantle ____ toward the ____ .

Section 17.4 Causes of Plate Motions (continued)

⟨Main Idea⟩ ———— **⟨Details⟩** ——————————————————————

Mantle Convection

Use with page 487.

Create *a diagram to illustrate convection currents in the mantle. Use Figure 17.21 to help you.*

- convection current
- asthenosphere
- upper mantle
- trench
- crust
- convergent plate boundary
- slab pull
- divergent plate boundary
- ridge push
- lithosphere

Use with page 468.

Identify *unanswered questions that scientists still have about convection currents and plate tectonics.*

_____

_____

_____

_____

_____

ANALOGY Draw a connection between heating soup on the stove and convection currents in the Earth.

_____

_____

_____

_____

_____

Plate Tectonics Chapter Wrap-Up

In the "What I Wanted to Find Out" column, copy the questions you listed in the Chapter Preview. In the "What I Learned" column, write down the answers you discovered as you worked through the chapter.

W What I Wanted to Find Out	L What I Learned
1. _____ _____	1. _____ _____
2. _____ _____	2. _____ _____
3. _____ _____	3. _____ _____

Review

Use this checklist to help you study.

☐ Study your Science Notebook for this chapter.

☐ Study the definitions of vocabulary words.

☐ Review daily homework assignments.

☐ Reread the chapter and review the tables, graphs, and illustrations.

☐ Review the Section Assessment questions at the end of each section.

☐ Look over the Study Guide at the end of the chapter.

SUMMARIZE After reading this chapter, list three things you have learned about plate tectonics.

_____

_____

_____

Volcanism

Chapter Preview

Before you read the chapter, use the "What I Know" column to list three things you know about volcanoes. Then list three questions you have about volcanoes in the "What I Want to Find Out" column.

K What I Know	W What I Want to Find Out
1._____ _____	1._____ _____
2._____ _____	2._____ _____
3._____ _____	3._____ _____

Science Journal

Think about the Launch Lab you did to model magma movement and record your response in this science journal.

Describe what happened to the oil when it entered the beaker.

_____

_____

Describe what happened when salt was added to the oil.

_____

_____

Volcanism
Section 18.1 Volcanoes

Main Idea	**Details**
	Scan *the photos and read the captions in this section. Write two questions you think may be answered in this section.*
	1. _____
	2. _____
Review Vocabulary	*Use your text to define the following term.*
convergent boundary	_____ _____
New Vocabulary	*Use your text to define each term.*
volcanism	_____
hot spot	_____
flood basalt	_____
fissure	_____
conduit	_____
vent	_____
crater	_____
caldera	_____
shield volcano	_____
cinder cone	_____
composite volcano	_____

Section 18.1 Volcanoes (continued)

Main Idea	Details

Anatomy of a Volcano

Use with page 505.

Create *a cross section of a volcano. Show the interior of the volcano. Label the following:*

• crater

• magma chamber

• vent

Sequence *the steps in the formation of a lake in a caldera. Refer to Figure 18.8 for help. The first one has been completed for you.*

_____ Steam causes explosions.

_____ A volcano erupts many times.

_____ The caldera that is formed fills with water to form a lake.

_____ The top of the partially empty magma chamber collapses.

Use with page 506.

Write *each of the following terms on the line below its corresponding image.*

• Cinder-cone volcano   • Composite volcano

• Shield volcano

_____   _____

_____

Section 18.1 Volcanoes (continued)

Main Idea ——— **Details** ————————————————————

Use with page 506. **Compare and Contrast** *the three types of volcanoes by completing the table below.*

Type of volcano	Description	How does it form?	How explosive is its eruption?
Cinder-cone	steep sides, generally small		more explosive than shield
Shield			
Composite	larger, with steep slopes that are concave		

Using *knowledge from the previous chapter, where would you predict each the three types of volcanoes would form?*

_____

_____

_____

_____

Volcanism
Section 18.1 Volcanoes

Main Idea ——— | **Details** ————————————————

Types of Volcanoes

Use with page 507.

Go *to a map and name three types of shield and composite volcanoes. Refer to the previous question to explain why these volcanoes formed in their present location.*

_____

Contrast *convergent and divergent volcanism.*

_____

_____

_____

_____

_____

_____

Hot Spots

Use with page 503.

Complete *the following sentences to better understand hot spots.*

_____ are unusually hot regions of Earth's mantle. In these places, high-temperature plumes of mantle material _____ toward the surface. The heat of the plumes _____ rock into _____. This _____ melts through the _____ and forms _____.

REAL-WORLD CONNECTION

Yellowstone National Park sits atop a caldera formed 650,000 years ago. Describe evidence showing that this area is still a geologic hot spot.

_____

_____

_____

Section 18.2 Eruptions (continued)

Main Idea	Details

Scan *Section 1 of your text. Use the checklist below as a guide.*

- Read all section titles.
- Read all bold words.
- Read all tables and graphs.
- Look at all the pictures and read their captions.
- Think about what you already know about volcanoes.

Write three facts you discovered about magma.

1. _____

2. _____

3. _____

Review Vocabulary

Use the text to define the following term.

basaltic

_____

_____

New Vocabulary

Use the text to define the following term.

viscosity _____

tephra _____

pyroclastic flow _____

Academic Vocabulary

Define the following term.

aid _____

_____

Section 18.2 Eruptions (continued)

⟨Main Idea⟩ ——— **⟨Details⟩** ———————————————

How Magma Forms

Use with page 508.

List *three factors that affect the formation of magma.*

1. _____

2. _____

3. _____

Analyze *Figure 18.1 to complete the table. Indicate in the table whether temperature, pressure, and melting point increase or decrease for each of the conditions.*

	Pressure	Temperature	Melting Point of Wet Albite
Increasing Depth Below Earth's Surface			
Decreasing Depth Below Earth's Surface			

Illustrate *the relationship between the melting point of a wet rock and a dry piece of the same rock under the same amount of pressure by sketching a thermometer and marking a possible temperature for each type of rock.*

SYNTHESIZE If you were the engineer on an oil-drilling expedition, explain how you might use a graph such as the one in Figure 18.2 in your text.

_____

_____

_____

Section 18.3 Intrusive Activity

| ⟨Main Idea⟩ | ⟨Details⟩ |

Types of Magma

Use with page 510.

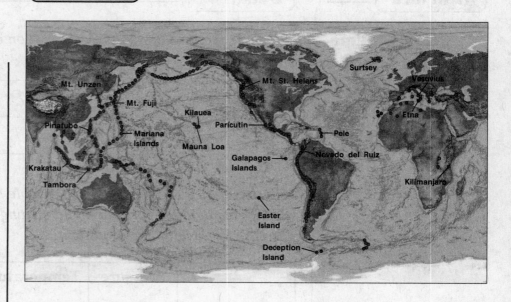

Identify *types of volcanoes. Write basaltic, andesitic, or rhyolithic to the left of each description.*

_____ Volcanoes that are found along continental margins. In the figure above, circle two such volcanoes in red.

_____ Volcanoes in which rocks in the upper mantle melt. In the figure above, circle two such volcanoes in blue.

Viscosity

Use with page 509.

Compare *the explosiveness, viscosity, and gas content of Surtsey to Tambora.*

_____

_____

_____

Organize *information about types of magma by completing the table below.*

		Andesetic Magma	
Gas content			4-6%
Silica Content	about 50%		about 10%
Viscosity		intermediate	
Explosives			

Volcanism

Section 18.3 Intrusive Activity (continued)

(**Main Idea**)————	(**Details**)———————————————
	Skim *Section 2 of your text. Write three questions that come to mind from reading the headings and the illustration captions.*

1. _____

2. _____

3. _____

Review Vocabulary

Use your text to define the following term.

igneous rock _____

New Vocabulary

In the left column, write the terms defined below.

_____ *intrusive igneous rock bodies*

_____ *irregularly shaped plutons that are similar to batholiths but smaller*

_____ *a pluton that forms when magma intrudes parallel to layers of rock*

_____ *the largest plutons*

_____ *a mushroom-shaped pluton with a round top and a flat bottom*

_____ *a pluton that cuts across preexisting rocks*

Section 18.3 Intrusive Activity (continued)

⊂**Main Idea**⊃ —————— ⊂**Details**⊃ ——————————————

List *three ways intruding magma can affect Earth's crust.*

1. _____

_____

2. _____

_____

3. _____

_____

Plutons

Use with page 514.

Record *the three characteristics used to classify plutons.*

1. _____

2. _____

3. _____

Draw *an illustration that includes the following features. Label the features on your drawing.*

- a batholith
- a dike
- a laccolith

- a sill
- a stock

Section 18.3 Intrusive Activity (continued)

(Main Idea) —— **(Details)** ———————————

Compare *a sill and a dike. Place each characteristic below in the Venn diagram to show whether it is a characteristic of a sill, a dike, or both.*

- cuts across preexisting rocks
- is parallel to the rocks it intrudes
- is a few centimeters to hundreds of meters thick
- many are coarse grained
- is a few centimeters to several meters wide
- is a pluton

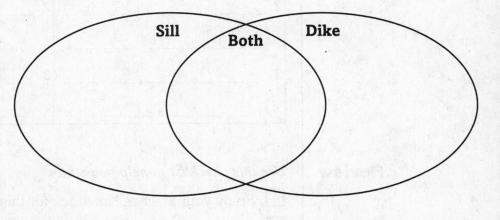

Plutons and Tectonics

Use with page 517.

Sequence *the four steps involved in forming batholiths from mountain-building processes. The first step has been completed for you.*

Two continental plates converge,
forcing continental crust into the mantle.

OR

Two oceanic plates converge and
one plate is subducted into the mantle.

Volcanism Chapter Wrap-Up

In the "What I Wanted to Find Out" column, copy the questions you listed in the Chapter Preview. In the "What I Learned" column, write down the answers you discovered as you worked through the chapter.

W What I Wanted to Find Out	L What I Learned
1. _____ _____	1. _____ _____
2. _____ _____	2. _____ _____
3. _____ _____	3. _____ _____

Review *Use this checklist to help you study.*

☐ Study your Science Notebook for this chapter.

☐ Study the definitions of vocabulary words.

☐ Review daily homework assignments.

☐ Reread the chapter and review the tables, graphs, and illustrations.

☐ Review the Section Assessment questions at the end of each section.

☐ Look over the Study Guide at the end of the chapter.

SUMMARIZE After reading this chapter, list three things you have learned about volcanic activity.

_____

_____

_____

Earthquakes
Chapter Preview

Before you read the chapter, use the "What I Know" column to list three things you know about earthquakes. Then list three questions you have about earthquakes in the "What I Want to Find Out" column.

K What I Know	W What I Want to Find Out
1. _____ _____	1. _____
2. _____ _____	2. _____
3. _____ _____	3. _____ _____

Science Journal

Think about the Launch Lab you did to model an earthquake and record your response in this science journal.

Compare the two movements.

_____

_____

_____

Infer which of the two scenarios accurately models what happens during an earthquake.

_____

_____

_____

Earthquakes

Section 19.1 Forces Within Earth

Main Idea —————— **Details** ————————————

Consider *the title of Section 1. List three topics that might be discussed in this section.*

1. _____

2. _____

3. _____

Review Vocabulary   *Use your text to define the following term.*

fracture _____

New Vocabulary   *Use your text to define each term.*

stress _____

strain _____

elastic deformation _____

plastic deformation _____

_____

fault _____

_____

seismic wave _____

primary wave _____

_____

secondary wave _____

_____

focus _____

_____

epicenter _____

Section 19.1 Forces Within Earth (continued)

⟨ **Main Idea** ⟩————   ⟨ **Details** ⟩————————————————————————

Stress and Strain

Use with pages 528–529.

Identify *the three kinds of stress that can act on Earth's rocks. Explain how each type of stress affects rocks.*

1. _____

2. _____

3. _____

Describe *what happens to a rubber band as it is stretched at each point or segment on the stress-strain curve below.*

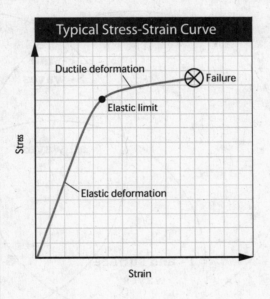

Typical Stress-Strain Curve

Ductile deformation

Failure

Elastic limit

Stress

Elastic deformation

Strain

Faults

Use with pages 530–531.

Compare *the different types of faults, and their characteristics by completing the table below.*

Type of Fault			
Cause		horizontal tension	
Movement			

Section 19.1 Forces Within Earth (continued)

Main Idea ————   **Details** ———————————————————

Earthquake Waves

Use pages 532–533.

Differentiate *between P-waves, S-waves, surface waves, both P- and S-waves, both S- and surface waves, or all three waves. Place each of the following characteristic in the Venn diagram below.*

- are body waves
- move up and down
- are seismic waves
- pass through Earth's interior
- cause rocks to move at right angles to the wave

- squeeze and pull rocks in the same direction as the waves
- move back and forth
- travel along Earth's surface
- move in two directions as they pass through rock

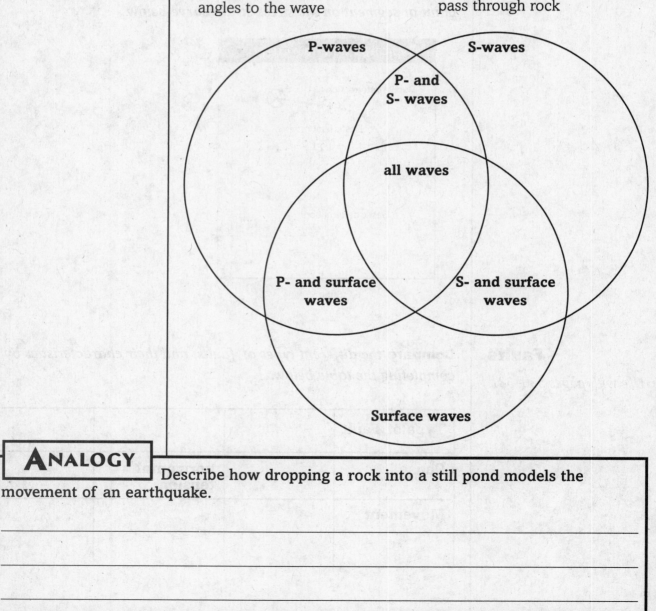

ANALOGY   Describe how dropping a rock into a still pond models the movement of an earthquake.

_____

_____

_____

Earthquakes

Section 19.2 Seismic Waves and Earth's Interior

Main Idea ——— **Details** ————————————————————————

Scan *Section 2 of your text. Use the checklist below as a guide.*

• Read all section titles.

• Read all bold words.

• Read all tables and graphs.

• Look at all the pictures and read their captions.

• Think about what you already know about seismic waves.

Write three facts you discovered about seismic waves as you scanned the section.

1. _____

2. _____

3. _____

Review Vocabulary *Use your text to define the following term.*

mantle _____

_____

New Vocabulary *Use your text to define each term.*

seismometer _____

_____

seismogram _____

_____

Academic Vocabulary *Define the following term.*

encounter _____

_____

Section 19.2 Seismic Waves and Earth's Interior (continued)

◁**Main Idea**▷ ———— ◁**Details**▷ ——————————————

Seismometers and Seismograms

Use with pages 534–535.

Label *the parts of the seismometer below.*

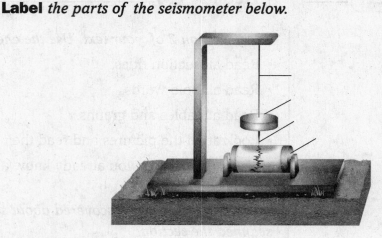

What if *a younger student is looking at the diagram above with you? Explain to the student how the seismometer works.*

_____

_____

_____

_____

_____

_____

Create *two questions that can be answered using information from Figure 19.9 in your book. For example, "How long did it take the S-waves to move 2000 km from the epicenter of the earthquake?"*

1._____

_____

_____

2._____

_____

_____

Section 19.2 Seismic Waves and Earth's Interior (continued)

Main Idea ————

Clues to Earth's Interior

Use with page 536.

Details ————

Sketch *a model of the interior of Earth. Label the following:*

- inner core
- mantle
- outer core

Once you have drawn your model, draw an earthquake focus on the left side of your model. Then add the following:

- P-waves
- P-wave shadow zones
- S-waves

Identify on your model where there are no direct P-waves and no direct S-waves.

SYNTHESIZE What would happen if S-waves encountered a lake or pond?
Explain your reasoning.

_____

_____

_____

Earthquakes

Section 19.3 Measuring and Locating Earthquakes

(Main Idea) _____ **(Details)** _____

Consider *the objectives on the first page of Section 3. Change each objective into a question that will be answered with information in the section.*

1. _____

    _____

2. _____

    _____

3. _____

(New Vocabulary) *Use the terms in the left margin to complete the following sentences.*

magnitude

amplitude

Richter scale

moment magnitude scale

modified Mercalli scale

When an earthquake is reported on the news, reporters often refer

to its _____ . This is the amount of energy released during an

earthquake. There is more than one way to measure it. _____

indicates the size of a seismic wave. One type of scale, the

_____ , measures the amount of damage from an

earthquake. The _____ accounts for the magni-

tude, size of the fault rupture, amount of movement along the fault,

and the rocks' stiffness. The _____ is based on the largest

seismic waves made by an earthquake.

Section 19.3 Measuring and Locating Earthquakes (continued)

Main Idea ————— **Details** ——————————————————

Earthquake Magnitude and Intensity

Use with pages 539–541.

Differentiate *between the three scales that are used to describe earthquakes. Write the name of each scale and describe what it measures in the concept map.*

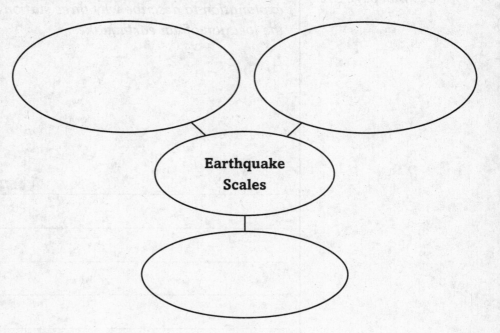

Earthquake Scales

Describe *the difference between the magnitude and the intensity of an earthquake.*

_____

_____

_____

_____

State *the three ways the depth of an earthquake's focus can be classified.*

1. _____

2. _____

3. _____

From the three classifications above, circle the depth of focus for almost all catastrophic quakes.

Section 19.3 Measuring and Locating Earthquakes (continued)

Main Idea	Details

Locating an Earthquake

Use with page 542.

Consider *the locations of two seismic stations as seen in Figure 19.17. Explain why it is impossible to determine the location of an earthquake using only two seismic stations. Then continue your explanation to describe why three stations are needed to determine the location of an earthquake.*

_____

_____

_____

_____

_____

Seismic Belts

Use with pages 543–544.

Describe *seismic belts by completing the paragraph below.*

Most of the world's earthquakes occur in narrow _____

that separate large regions of little or no _____ . Most

earthquakes are associated with _____ . A

_____ percentage of earthquakes happen far from _____

_____ and are distributed more or less at _____ .

ANALOGY Point out how you could use a simple puzzle to demonstrate where most earthquakes occur in the world. Describe what would happen if you shook the puzzle.

_____

_____

_____

Earthquakes
Section 19.4 Earthquakes and Society

Main Idea | **Details**

Skim *Section 4 of your text. Write three questions that come to mind from reading the headings and the illustration captions.*

1. _____

2. _____

3. _____

Review Vocabulary

Use your text to define the following term.

geology

_____

_____

New Vocabulary

Use your text to define each term.

tsunami

_____

_____

soil liquefaction

seismic gap

_____

_____

Show your understanding of each of the above terms by using each one in a sentence of your own.

_____

_____

_____

Section 19.4 Earthquakes and Society (continued)

Main Idea ———— **Details** ————————————————————

Some Earthquake Hazards

Use with pages 545–548.

Compare *Place each type of building on the continuum to show its strength and quality during an earthquake.*

- buildings on large rubber structures
- high-rise, steel-frame buildings
- unreinforced buildings of stone or concrete
- wooden structures

Least amount of earthquake damage   **Greatest amount of earthquake damage**

←——————|————————|————————|————————|——————→

Outline *information about earthquake hazards. Describe each hazard in detail.*

   I. The 4 types of earthquake hazards are:

      _____

      A. _____

         1. Pancaking—_____

            _____

         2. Failure related to height—_____

            _____

            _____

      B. _____

         1. Landslides—_____

         2. _____

            _____

         3. _____

            _____

      C. _____—is an area of great vertical offset where the fault intersects the ground surface

      D. _____

         _____

Section 19.4 Earthquakes and Society (continued)

Main Idea	Details

Seismic Risk

Use with page 549.

Identify *four states that have the lowest seismic risk. Refer to the seismic risk map of the United States on page 549 of your text.*

1. _____

2. _____

3. _____

4. _____

Infer *whether any of these states will ever have an earthquake.*

_____

_____

_____

Earthquake Prediction

Use with pages 549–551.

Organize *information about earthquake prediction by completing the graphic organizer below.*

**Earthquake Prediction
is based on**

REAL-WORLD CONNECTION

Imagine you live in an area that has had three minor earthquakes and one major earthquake in the last 100 years. It has been 40 years since the last earthquake. Assess the probability of another earthquake in your area.

_____

_____

Earthquakes Chapter Wrap-Up

In the "What I Wanted to Find Out" column, copy the questions you listed in the Chapter Preview. In the "What I Learned" column, write down the answers you discovered as you worked through the chapter.

W What I Wanted to Find Out	L What I Learned
1. _____ _____	1. _____ _____
2. _____ _____	2. _____ _____
3. _____ _____	3. _____ _____

Review

Use this checklist to help you study.

☐ Study your Science Notebook for this chapter.

☐ Study the definitions of vocabulary words.

☐ Review daily homework assignments.

☐ Reread the chapter and review the tables, graphs, and illustrations.

☐ Review the Section Assessment questions at the end of each section.

☐ Look over the Study Guide at the end of the chapter.

SUMMARIZE *After reading this chapter, list three things you have learned* about earthquakes.

_____

_____

_____

Mountain Building
Chapter Preview

Before you read the chapter, use the "What I Know" column to list three things you know about mountains. Then list three questions you have about mountains in the "What I Want to Find Out" column.

K What I Know	W What I Want to Find Out
1. _____ _____	1. _____ _____
2. _____ _____	2. _____ _____
3. _____ _____	3. _____ _____

Science Journal

Think about the Launch Lab you did to model crustal differences and record your response in this science journal.

How does density affect the height of flotation?

_____

_____

How does thickness affect the height of flotation?

_____

_____

Which block represents oceanic crust? Continental crust?

_____

_____

Mountain Building
Section 20.1 Crust-Mantle Relationships

Main Idea ——— **Details** ———

Consider *the objectives on the first page of Section 1. Write three questions that come to mind while reading these statements.*

1. _____

2. _____

3. _____

Review Vocabulary   *Use your text to define the following term.*

equilibrium   _____

_____

New Vocabulary   *Use your text to define each term.*

topography   _____

isostasy   _____

_____

root   _____

isostatic rebound   _____

Academic Vocabulary   *Define the following term.*

mode   _____

_____

Section 20.1 Crust-Mantle Relationships (continued)

Main Idea	Details

Earth's Topography

Use with page 562.

Identify *the two modes that most of Earth's elevations cluster around. Then classify those elevations as oceanic crust or continental crust.*

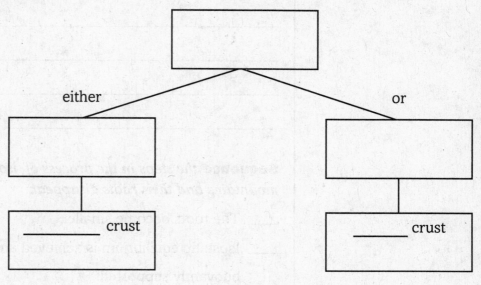

either   or

_____ crust   _____ crust

Name *the two main factors that lead to differences in elevation in Earth's crust. Describe how each affects elevation.*

1._____

_____

2._____

_____

Infer *how the elevations of oceanic crust would be different if oceanic crust were thicker. Explain.*

_____

_____

_____

_____

Section 20.1 Crust-Mantle Relationships (continued)

Main Idea ———— **Details** —————————————————

Isostasy

Use with pages 563–564.

Consider *the principle of isostasy and how it relates to mountains. Describe what happens to the roots of a mountain range as the mountain range is forming.*

_____

_____

_____

_____

Sequence *the steps in the process of isostatic rebound until mountains and their roots disappear.*

_____ The roots become smaller.

_____ Isostatic equilibrium is achieved and the mountains are buoyantly supported.

_____ This process continues until the mountains and their roots disappear.

_____ Mountain peaks erode, decreasing the mass of the mountains.

_____ Mountains rise above Earth's surface, forming deep roots.

ANALOGY Imagine a cargo boat loaded with grain is being unloaded in a harbor. Use this situation to illustrate isostatic rebound. Draw a series of images of the boat and explain how what is happening as it is unloaded is similar to erosion.

Mountain Building
Section 20.2 Orogeny

Main Idea ———— **Details** ————————————————

Skim *Section 2 of your text. Write three questions that come to mind while reading the headings and the illustration captions.*

1. _____

2. _____

3. _____

Review Vocabulary *Use your text to define the following term.*

island arc
_____
_____
_____

New Vocabulary *Use your text to define the following term.*

orogeny
_____
_____
compressive forces
_____
_____

Show your understanding of the term by using it in a sentence.
_____
_____
_____

Academic Vocabulary *Define the following term.*

complex
_____
_____

Section 20.2 Orogeny (continued)

| Main Idea | Details |

Orogeny

Use with pages 567–570.

Complete *the graphic organizer about orogeny.*

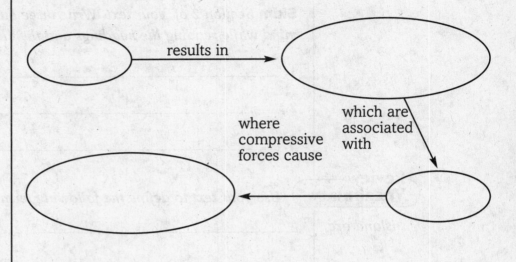

Compare *the different types of mountains that form along convergent plate boundaries in the table below.*

Type of Boundary	Types of Mountains that Form	How the Mountains are Formed	Type of Rocks Associated with Mountains

Section 20.2 Orogeny (continued)

⟨**Main Idea**⟩———— ⟨**Details**⟩————————————

The Appalachian Mountains—A Case Study

Use with page 571.

Sequence *the formation of the Appalachian Mountains. Some of the steps have been completed for you. In the smaller boxes, write the name of the region of the Appalachians that was formed at that step.*

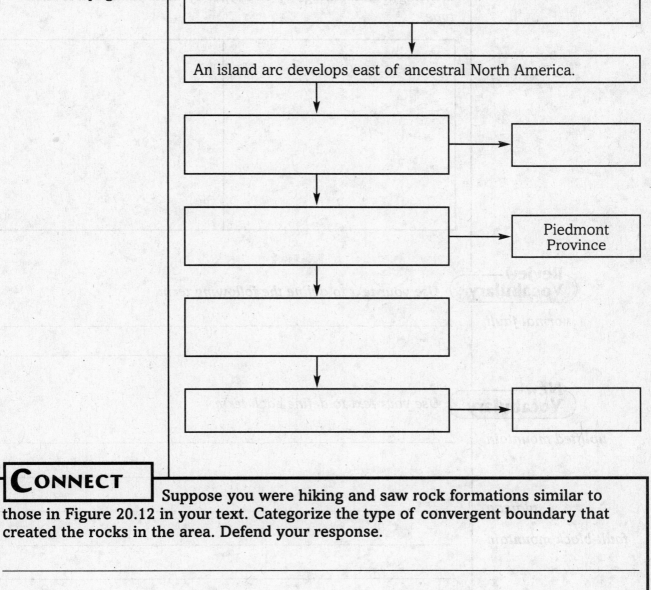

An island arc develops east of ancestral North America.

Piedmont Province

⟨**CONNECT**⟩

Suppose you were hiking and saw rock formations similar to those in Figure 20.12 in your text. Categorize the type of convergent boundary that created the rocks in the area. Defend your response.

_____

_____

_____

_____

_____

Mountain Building
Section 20.3 Other Types of Mountains

Main Idea _____

Details

Examine *the photographs of different types of mountains in Section 3 of your text. Use the photos to create an informative drawing with an image of two types of mountains. Label each type.*

Review Vocabulary

Use your text to define the following term.

normal fault _____

_____

New Vocabulary

Use your text to define each term.

uplifted mountain _____

_____

plateau _____

fault-block mountain _____

_____

Academic Vocabulary

Define the following term.

dynamic _____

_____

Section 20.3 Other Types of Mountains (continued)

<Main Idea>——— <Details>————————————————————

Divergent-Boundary Mountains

Use with pages 572–573.

Illustrate *how an ocean ridge forms. Show the following:*

- central rift
- lithosphere
- magma upper
- mantle
- new crust
- older crust

Draw an arrow to show the direction that the magma and crust move in relation to the central rift. Refer to your model to describe the process of forming an ocean ridge.

_____

_____

_____

_____

_____

Illustrate *vertical dikes and pillow basalts, which are characteristic of ocean-ridge rocks. Label each and write a description of how each is formed.*

_____

_____

_____

_____

_____

Section 20.3 Other Types of Mountains (continued)

Main Idea —— **Details** ————————————————————

Uplifted Mountains

Use with pages 575–576.

Compare and Contrast *uplifted and fault-block mountains in the Venn diagram. Place each characteristic in the diagram to show whether it is a trait of uplifted mountains, fault-block mountains, or both.*

- crust is uplifted as a unit
- form along faults
- have little structural deformation
- formed by internal forces
- pieces of crust are tilted, uplifted, or dropped

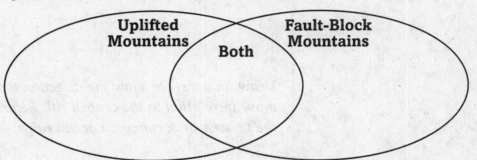

Uplifted Mountains   Both   Fault-Block Mountains

Appalachian Case Study

Use with pages 571–573.

Sequence *the steps in the formation of the Appalachian Mountains by placing these in the correct order.*

- An island arc forms between Africa and North America.
- The continental edge becomes a folded valley and ridge region as the fragment and arc attach to North America.
- A continental fragment attaches to North America and becomes the blue ridge region.
- North America and Africa divide.
- The island arc attaches to North America to become the piedmont region.

1. _____

2. _____

3. _____

4. _____

5. _____

SYNTHESIZE Recall what you learned in Chapter 18 about different types of volcanoes. Infer whether volcanoes that form over divergent boundaries are more or less explosive than most volcanoes that form along plate boundaries. Explain your reasoning.

_____

_____

_____

Tie-It-All-Together

What if Earth was not a dynamic planet, but had stayed the same for millions of years? What might Earth's surface look like today?

_____

_____

_____

_____

_____

Compare mountains that form at plate boundaries to nonboundary mountains.

_____

_____

_____

_____

Design a model of one of the types of mountains in the chapter. It should be a working model that can show the formation of the type of mountain. Be creative, yet think about constructing a simple model. Sketch your model below, then build it. Demonstrate your model to your class.

Mountain Building Chapter Wrap-Up

In the "What I Wanted to Find Out" column, copy the questions you listed in the Chapter Preview. In the "What I Learned" column, write down the answers you discovered as you worked through the chapter.

W What I Wanted to Find Out	L What I Learned
1._____ _____	1._____ _____
2._____ _____	2._____ _____
3._____ _____	3._____ _____

Review   *Use this checklist to help you study.*

☐ Study your Science Notebook for this chapter.

☐ Study the definitions of vocabulary words.

☐ Review daily homework assignments.

☐ Reread the chapter and review the tables, graphs, and illustrations.

☐ Review the Section Assessment questions at the end of each section.

☐ Look over the Study Guide at the end of the chapter.

SUMMARIZE   After reading this chapter, state three things you have learned about how mountains are formed.

_____

_____

_____

Fossils and the Rock Record
Chapter Preview

Before you read the chapter, use the "What I Know" column to list three things you know about fossils. Then list three questions you have about fossils in the "What I Want to Find Out" column.

K What I Know	W What I Want to Find Out
1. _____ _____	1. _____ _____
2. _____ _____	2. _____ _____
3. _____ _____	3. _____ _____

Science Journal

Think about the Launch Lab you did to model a fossil hunt, and record your response in this science journal.

Describe the appearance of the spongue upon digging it up.

_____

_____

_____

Explain how the spongue in this experiment is similar to a fossil.

_____

_____

_____

Fossils and the Rock Record

Section 21.1 The Rock Record

Main Idea

Details

Examine *the geologic time scale in Figure 21.1 on page 554 of your text. Write three things about the geologic time scale that interest you.*

1. _____

2. _____

3. _____

New Vocabulary

Use your text to define each term.

geologic time scale

_____

_____

_____

eon

_____

_____

Precambrian

_____

era

_____

_____

period

_____

_____

epoch

_____

_____

mass extinction

_____

_____

Academic Vocabulary

Define the following term.

decline

_____

Section 21.1 The Rock Record (continued)

Main Idea	Details

The Rock Record

Use with pages 590–591.

Identify *what scientists use to divide the history of Earth into time units.*

_____

List *the three main factors that scientists consider when refining the geologic time scale.*

1. _____

2. _____

3. _____

Geologic Time

Use with pages 591-593

Construct *three quiz questions using information from the time scale in Figure 21.1. Then write the answers to the questions.*

Question: _____

_____

Answer: _____

Question: _____

_____

Answer: _____

Question: _____

_____

Answer: _____

Label eon, era, epoch, *and* period *in the correct location on the continuum.*

Shortest length of time   Longest length of time

←――――――――――――――――――――――――――――→

Section 21.1 The Rock Record (continued)

⸨Main Idea⸩ —————— ⸨Details⸩ ——————————————————————

Main Ideas

Use with page 591.

Label *the simplified geologic time scale below. Use Figure 21.2 to help you. Include the different eons and eras on your time scale. On the lines to the left and right, write important events about the appearance and development of different animals.*

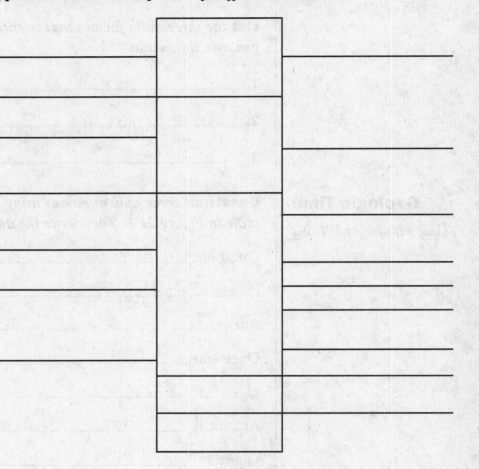

⌈**C**RITIQUE⌉ Critique the following statement: "Humans are relatively old according to the geologic time scale." Explain whether you agree or disagree. Defend your answer.

_____

_____

_____

_____

_____

Fossils and the Rock Record

Section 21.2 Relative-Age Dating

Main Idea ———

Details ——————————

Scan *Section 2 of your text. Use the checklist below as a guide.*

• Read all section titles.

• Read all bold words.

• Read all tables and graphs.

• Look at all the pictures and read their captions.

• Think about what you already know about the dating of rocks.

Write two facts you discovered about relative-age dating of rocks.

1. _____

2. _____

Review Vocabulary

Use your text to define the following term.

granite _____

_____

New Vocabulary

Use your text to define each term.

uniformitarianism _____

relative-age dating _____

original horizontality _____

_____

superposition _____

_____

cross-cutting relationships _____

principle of inclusions _____

unconformity _____

correlation _____

key bed _____

Section 21.2 Relative-Age Dating (continued)

Main Idea ——— **Details** —————————————————

Principles for Determining Relative Age

Use with pages 595-597.

Identify *the three geologic principles for determining the relative age of rocks.*

1._____

2._____

3._____

Draw *two horizontal layers of rock. Then draw and label inclusions in the top layer. Describe how inclusions are formed.*

_____

Create *a diagram of four or five rock layers. Include one layer with inclusion. Draw an intrusion that cuts through more than one layer. Label the layers A, B, C,.... Write three true statements about the relative ages of the rock layers in your diagram. State the geologic principle that helped you write each statement.*

1._____

2._____

3._____

Section 21.2 Relative-Age Dating (continued)

⸨Main Idea⸩———— ⸨Details⸩————————————

Unconformities

Use with pages 598.

Compare *the different types of unconformities using the graphic organizer.*

Type:

How formed?

Type:

How formed?

Uncomformities

Type:

How formed?

⸨**S**YNTHESIZE⸩ **Hypothesize** *what could have happened to separate the Permian Kaibab Formation that rims the top of the Grand Canyon and the Permian Kaibab Formation that is found 300 km away at the bottom of a 200-m gorge.*

_____

_____

_____

_____

_____

Fossils and the Rock Record
Section 21.3 Absolute-Age Dating

Main Idea   **Details**

Skim *Section 3 of your text. Read the headings and captions more closely. Write three questions you think may be answered in this section.*

1. _____

2. _____

3. _____

Review Vocabulary   *Use your text to define the following term.*

isotope   _____

New Vocabulary   *Use your text to define each term.*

absolute-age dating   _____

_____

radioactive decay   _____

_____

radiometric dating   _____

_____

half-life   _____

_____

radiocarbon dating   _____

dendochronology   _____

_____

varve   _____

_____

_____

Section 21.3 Absolute-Age Dating (continued)

Main Idea	Details

Use with page 601.

Compare *the relative-age dating of rocks to absolute-age dating of rocks using the graphic organizer*

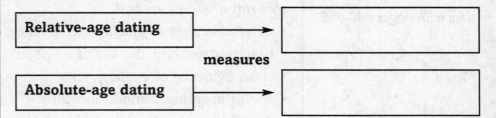

Relative-age dating	⟶	

measures

Absolute-age dating	⟶	

Use of Radioactive Isotopes

Use with pages 601-603.

Create *an illustration to explain the concept of half-life. Use one of the radioactive isotopes listed in Table 21.1. Your illustration should show the amount of parent element at the beginning and the amount of parent and daughter elements after one half-life.*

Explain *the radioactive decay of uranium-238 to lead-206 in the table below. Use Table 21.1 in your book for help.*

Radioactive Decay of Uranium-238 to Lead-206				
	Percent Parent Element	Percent Daughter Element	Elapsed Years	Number of Half-Lives
Time 1				0
Time 2				1
Time 3				2
Time 4				3
Time 5				4

Section 21.3 Absolute-Age Dating (continued)

| Main Idea | Details |

Other Ways to Determine Age

Use with pages 604-605.

Compare and Contrast *tree rings and varves. Place each characteristic in the Venn diagram to show whether it is a trait of tree rings, varves, or both.*

- determine ages of glacial lake sediments
- used to determine the age of an object
- show evidence of cyclic events
- widest in spring, thinnest in winter
- thickest in summer, thinnest in winter

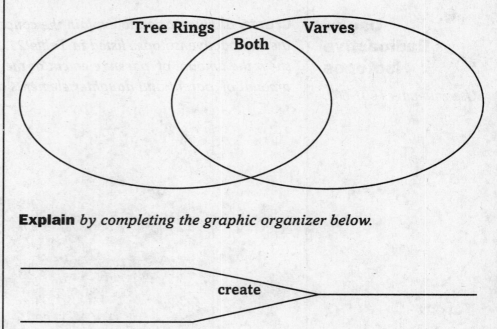

Tree Rings   Both   Varves

Explain *by completing the graphic organizer below.*

create

SYNTHESIZE *Draw a cross section of a tree that is six years old. The tree lives in a normal environment. Next to it, create a cross section of a tree that is six years old, but has lived through six years of drought. Explain why the two drawings look different from each other.*

Fossils and the Rock Record
Section 21.4 Fossil Remains

Main Idea _____

Details _____

Predict *three things that might be discussed in this section based on its title.*

1. _____

2. _____

3. _____

New Vocabulary *Use your text to define each term.*

evolution _____

original preservation _____

_____

altered hard part _____

_____

mineral replacement _____

_____

mold _____

_____

cast _____

_____

trace fossil _____

_____

index fossil _____

_____

_____

Section 21.4 Fossil Remains (continued)

Main Idea

Details

Use with page 606.

Identify *three ways fossils are useful.*

1. _____

2. _____

3. _____

_____

Types of Fossils

Use with pages 606-609.

Compare *types of fossils by completing the table below.*

Type of Fossil	Description and Information about Its Formation	Example
Original preservation		
		petrified wood
Index fossils		
		trilobite molds, casts of molds
Indirect evidence of past life		

Section 21.4 Fossil Remains (continued)

Main Idea ——— **Details** ———————————————————

Use with page 608.

Sequence *the steps in the process of forming a mold and a cast.*

___ The shell is buried.

___ Sediment fills the mold.

___ A hermit crab crawls out of its shell to find a bigger one.

___ A mold is left in the shell's place.

___ Groundwater dissolves the shell.

___ A cast of the shell is made.

Use with page 569.

Illustrate *how an index fossil can assist in relative age dating of rock units.*

REAL-WORLD CONNECTION Think about fossils could be made from organisms that are currently living. Describe two fossils you could create that would help future generations learn about how you lived. Be sure to include details about how the fossils might form.

_____

_____

_____

_____

Fossils and The Rock Record Chapter Wrap-Up

In the "What I Wanted to Find Out" column, copy the questions you listed in the Chapter Preview. In the "What I Learned" column, write down the answers you discovered as you worked through the chapter.

W What I Wanted to Find Out	L What I Learned
1. _____	1. _____
2. _____	2. _____
_____	_____
3. _____	3. _____
_____	_____
_____	_____

Review *Use this checklist to help you study.*

☐ Study your Science Notebook for this chapter.

☐ Study the definitions of vocabulary words.

☐ Review daily homework assignments.

☐ Reread the chapter and review the tables, graphs, and illustrations.

☐ Review the Section Assessment questions at the end of each section.

☐ Look over the Study Guide at the end of the chapter.

SUMMARIZE After reading this chapter, describe three things you have learned about fossils and the rock record.

_____

_____

_____

The Precambrian Earth

Chapter Preview

Before you read the chapter, use the "What I Know" column to list three things you know about the Precambrian Era. Then list three questions you have about the Precambrian Era in the "What I Want to Find Out" column.

K What I Know	W What I Want to Find Out
1. _____ _____	1. _____ _____
2. _____ _____	2. _____ _____
3. _____ _____	3. _____ _____

Science Journal

Observe

Describe what happened to the colored water and vegetable oil in the beaker.

_____

Explain how this is similar to what happened to the core and mantle when Earth formed.

_____

_____

_____

_____

_____

_____

The Precambrian Earth
Section 22.1 Early Earth

Main Idea ———— **Details** ————————————————

Predict *Read the title of Section 1. List three things that might be discussed in this section.*

1. _____

2. _____

3. _____

Review Vocabulary   *Use your text to define the following term.*

metamorphism _____

_____

New Vocabulary   *Use your text to define each term.*

zircon _____

_____

meteorite _____

_____

asteroid _____

_____

_____

Academic Vocabulary   *Define the following term.*

generate _____

Section 22.1 Early Earth (continued)

⟨**Main Idea**⟩ ── ⟨**Details**⟩ ────────────────────────

Earth's "Birth"

Use with page 620.

Organize *the four billion years that make up the Precambrian Era by completing the following flowchart.*

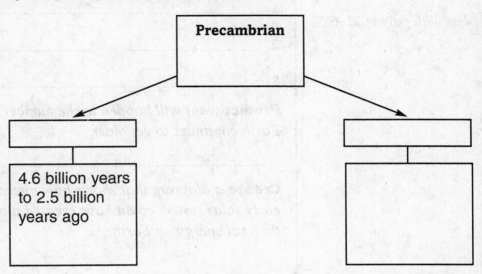

How Old Is Earth?

Use with pages 620–621.

Examine *the evidence scientists used to determine Earth's age by completing the table below. The first one has been done for you.*

Rock or Mineral	Age
rocks in Earth's crusts	3.96 to 3.8 billion years old
	4.1 to 4.2 billion years old
	between 4.5 and 4.7 billion years old
	approx 4.6 billion years old

Explain *how scientists used the evidence in the chart to determine the age of Earth and give their conclusion.*

_____

_____

_____

_____

Section 22.1 Early Earth (continued)

Main Idea ————   **Details** ————————————————

Earth's Heat Sources

Use with pages 621–622.

List *the three heat sources of Earth shortly after it formed.*

1. _____

2. _____

3. _____

Predict *what will happen to the number of radioactive isotopes as Earth continues to get older.*

_____

Create *a diagram that shows how meteorites and asteroids in the early solar system could have created a tremendous amount of thermal energy on Earth.*

Explain *gravitational contraction by completing the following sentences.*

Another source of Earth's heat was _____ .

Meteor bombardment and the resulting accumulation of meteorite

material caused Earth to _____ . The _____ of the material

caused _____ of the layers beneath. This energy

was converted into _____ . The meteorite's material also

caused a _____ , which kept the ____ from escaping.

REAL-WORLD CONNECTION The interior of Earth is still generating heat. Hypothesize which of the three original sources this heat is coming from.

_____

_____

_____

The Precambrian Earth
Section 22.2 Formation of the Crust and Continents

Main Idea ——— **Details** ————————————————————————

Skim *Section 2 of your text. Read the headings and the figure captions. Write three questions that come to mind.*

1. _____

2. _____

3. _____

Review Vocabulary *Use your text to define the following term.*

magma _____

New Vocabulary *Use your text to define each term.*

differentiation _____

_____

microcontinent _____

craton _____

Precambrian shield _____

Canadian Shield _____

Laurentia _____

Academic Vocabulary *Define the following term.*

illustrate _____

Section 22.2 Formation of the Crust and Continents (continued)

(**Main Idea**)——— (**Details**)————————————

Formation of the Crust

Use with pages 623–624.

Model *the differentiation of Earth. Draw a cross-section of Earth and label five layers. Use Figure 22.3 to help you.*

Analyze *differentiation and the formation of the crust in the concept web below.*

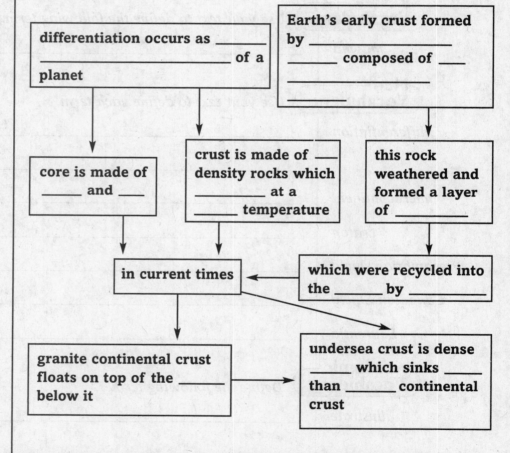

differentiation occurs as _____ _____ of a planet

Earth's early crust formed by _____ _____ composed of ____ _____ _____

core is made of _____ and ____

crust is made of ____ density rocks which _____ at a _____ temperature

this rock weathered and formed a layer of _____

in current times

which were recycled into the _____ by _____

granite continental crust floats on top of the _____ below it

undersea crust is dense _____ which sinks _____ than _____ continental crust

The Cores of the Continents

Use with page 625.

Explain *the cores of the continents by completing the following sentences.*

Continents contain a core of _____ and _____ rock called

a _____ . The Precambrian shield of North America is

called the _____ . It is exposed in parts of _____ ,

_____ , _____ , Michigan, _____ , and Greenland.

Section 22.2 Formation of the Crust and Continents (continued)

⟨Main Idea⟩———— ⟨Details⟩——————————————

Growth of Continents

Use with pages 625–627.

Create *cartoon-strip style sketches to show a chronology of the formation of Earth's surface as we now know it. Draw at least four frames. Include events such as:*

- collision of microcontinents
- formation of microcontinents
- Grenville orogeny
- suturing at orogenies to produce larger continents
- formation of Laurentia
- formation of Rodinia

List *four important developments of Earth by the end of the Proterozoic.*

1. _____

2. _____

3. _____

4. _____

┌───┐
│ **REAL-WORLD CONNECTION**
 Wegener mentioned Pangaea in his
 hypothesis about continental drift. Compare and contrast Pangaea and Rodinina.

 _____

 _____

 _____
└───┘

The Precambrian Earth

Section 22.3 Formation of the Atmosphere and Oceans

Main Idea | **Details**

Objectives *Read the objectives on the first page of Section 3. Write three questions that come to mind from reading these statements.*

1. _____

2. _____

3. _____

Review Vocabulary | *Use your text to define the following term. Hypothesize how the atmosphere in early times was different from now.*

ultraviolet radiation _____

_____

New Vocabulary | *Use your text to define each term.*

cyanobacteria _____

_____

stromatolite _____

_____

banded iron formation _____

_____

_____

red bed _____

Academic Vocabulary | *Define the following term.*

dominate _____

Section 22.3 Formation of the Atmosphere and Oceans (continued)

Main Idea	Details

The Precambrian Atmosphere

Use with page 628.

Summarize *the formation of seven gases in the early atmosphere.*

```
┌──────────────┐      ┌──────────────┐      ┌──────────────┐
│ Early        │ ───▶ │ decay of     │ ───▶ │              │
│ Atmosphere   │      │ radioactive  │      │              │
│              │      │ potassium-40 │      │              │
└──────┬───────┘      └──────────────┘      └──────────────┘
       │
       ▼                                     ┌──────────────┐
┌──────────────┐                             │              │
│ out-gassing  │ ─────────────────────────▶  │              │
│ by           │                             │              │
│              │                             │              │
└──────────────┘                             └──────┬───────┘
                                                    │
                                                    ▼
                                             ┌──────────────┐
                                             │ chemical     │
                                             │ reactions    │
                                             └──────┬───────┘
                                                    │
                                                    ▼
                                             ┌──────────────┐
                                             │              │
                                             │              │
                                             └──────────────┘
```

Oxygen in the Atmosphere

Use with page 629.

Predict *what would have happened to life on the planet if cyanobacteria had not evolved 3.46 billion years ago.*

_____

_____

_____

_____

Compare *the rocks related to oxygen in the atmosphere.*

Rock Type	Composition	Ages	Evidence of Oxygen?
			Cyanobacteria were the first organisms to use photosynthesis, which produced oxygen.
Banded iron formation			The iron found in rocks began to oxidize in the presence of oxygen.

Section 22.3 Formation of the Atmosphere and Oceans (continued)

Main Idea ——— **Details** ———————————————————

Importance of Oxygen

Use with page 631.

List *two reasons oxygen is vital to life on Earth.*

1. _____

2. _____

Formation of the Oceans

Use with page 631.

Create *a diagram to show where water on Earth came from and why the water in the seas is salty. Include both ice and steam sources in your diagram.*

SUMMARIZE

Summarize changes described in this section that occurred during early Earth that made it a place for life.

_____

_____

_____

The Precambrian Earth

Section 22.4 Early Life on Earth

⟨Main Idea⟩———— **⟨Details⟩**——————————————

Scan *Section 4 in your text. Read the bold words, headings, and figure captions. Write four facts about early life that you discovered as you scanned the section.*

1. _____

2. _____

3. _____

4. _____

⟨Review Vocabulary⟩ *Use your text to define the following term.*

hydrothermal vent _____

_____

⟨New Vocabulary⟩ *Use your text to define each term.*

amino acids _____

prokaryote _____

eukaryote _____

Ediacara biota _____

_____

⟨Academic Vocabulary⟩ *Define the following term.*

constantly _____

Section 22.4 Early Life on Earth (continued)

Main Idea	Details

Origin of Life on Earth

Use with pages 633–634.

Model *the "primordial soup" experiment by drawing Miller and Urey's laboratory setup. Label the two chambers, write the products found in the condensation chamber, the gases found in the atmospheric chamber, and the metal used for the electrodes.*

Proterozoic Life

Use with pages 635–636.

Organize *the information about DNA and RNA in the Venn diagram below.*

- carries instructions necessary for cells to function
- first replicating molecules on Earth
- need enzymes to replicate
- ribozymes replicate without enzymes

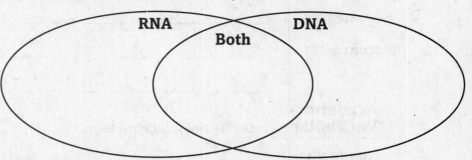

Compare and Contrast *the characteristics of early life forms by filling in the table below.*

Organism	Structure	Example	Approximate age
Prokaryote			
Eukaryote			

Section 22.4 Early Life on Earth (continued)

Main Idea ———— **Details** ————————————————————

Ediacara Fossils

Use with page 637.

Model *your own Ediacara animal. Be sure that the organism has all the features of Ediacaran fauna and that it lives in the appropriate location. Make your drawing in the space below. Label the different parts and describe how it lives.*

_____

_____

_____

Describe *Ediacara fossils by completing the following sentences.*

Some _____ fossils are similar to _____ , _____ ,

arthropods, and echinoderms. Some scientists believe that these

animals were _____ of the modern look-alikes. Other scientists

think that Ediacaran fauna do not lead to any modern animals.

They had no _____ , _____ , ___ , and probably did not _____ .

Ediacaran fossils have been found _____ .

SYNTHESIZE Tell where the Urey-Miller method of amino acid creation would
have had to occur. Describe how the same process could have happened at hydrothermal
vents and why scientists think life could have begun there.

_____

_____

_____

_____

The Precambrian Earth Chapter Wrap-Up

In the "What I Wanted to Find Out" column, copy the questions you listed in the Chapter Preview. In the "What I Learned" column, write down the answers you discovered as you worked through the chapter.

W What I Wanted to Find Out	L What I Learned
1. _____ _____	1. _____ _____
2. _____ _____	2. _____ _____
3. _____ _____	3. _____ _____

Review *Use this checklist to help you study.*

☐ Study your Science Notebook for this chapter.

☐ Study the definitions of vocabulary words.

☐ Review daily homework assignments.

☐ Reread the chapter and review the tables, graphs, and illustrations.

☐ Review the Section Assessment questions at the end of each section.

☐ Look over the Study Guide at the end of the chapter.

SUMMARIZE After reading this chapter, list three things you have learned about the Precambrian Earth.

_____

_____

_____

The Paleozoic, Mesozoic, and Cenozoic Eras
Chapter Preview

Before you read the chapter, use the "What I Know" column to list three things you know about the Paleozoic Era. Then list three questions you have about the Paleozoic Era in the "What I Want to Find Out" column.

K What I Know	W What I Want to Find Out
1. _____	1. _____
_____	_____
2. _____	2. _____
_____	_____
3. _____	3. _____
_____	_____

Science Journal

Think about the Launch Lab you did to observe how oil or water can be stored in solid rock and record your response in this science journal.

Sketch a cross-section of the rock or brick. Include sketches of both before and after you added the water to it. Include in your sketches what the inside of the brick might look like.

Infer what happened to the water.

_____

_____

_____

The Paleozoic, Mesozoic, and Cenozoic Eras
Section 23.1 The Paleozoic Era

Main Idea ————— **Details** ————————————————

Write *each objective from the beginning of the section in the form of a question. Answer your questions as you read your text.*

1. _____

_____

2. _____

_____

3. _____

_____

New Vocabulary *Use your text to define each term.*

paleogeography _____

passive margin _____

transgression _____

_____

regression _____

_____

Cambrian explosion _____

_____

Academic Vocabulary *Define the following term.*

indicate _____

Section 23.1 The Paleozoic Era (continued)

Main Idea ————   **Details** ———————————————————

Continental Setting

Use with page 651.

Analyze *Laurentia in the Cambrian Period by completing the table.*

Laurentia in the Cambrian Period	
Location	**Description**
Characteristic Sediments	**Tectonic Activity**

Changes in Sea Level

Use with page 649.

Define *the word "passive." Explain how it relates to the term "passive margin."*

_____

_____

Categorize *each characteristic as being a trait of a transgression or a regression.*

- deep-water deposits overlay shallow-water deposits
- shallow-water deposits overlay deep water deposits
- happens when sea level falls
- shoreline moves inland
- happens when sea level rises
- shoreline moves seaward

Transgression	**Regression**

Section 23.1 The Paleozoic Era (continued)

⟨**Main Idea**⟩ ——————   ⟨**Details**⟩ —————————————————————

Early Paleozoic Life

Use with page 649.

Draw *the cross-section of the shoreline of an ocean similar to Figure 23.2 on page 649. Label the original shoreline position. Then label the position of the shoreline after a transgression and the position of the shoreline after a regression.*

Cambrian Explosion

Use with page 653.

Describe *why the beginning of the Cambrian Period often is called the Cambrian "explosion."*

_____

_____

_____

_____

Identify *three facts about the fossilized Cambrian organisms that come from the Burgess Shale.*

1. _____

2. _____

3. _____

_____

SYNTHESIZE   Hypothesize what will happen to sediments that currently are being deposited if sea level were to rise.

_____

_____

_____

The Paleozoic, Mesozoic, and Cenozoic Eras
Section 23.2 The Mesozoic Era

Main Idea ———— **Details** ——————————————————

Skim *Section 1. Look at the headings, photos, illustrations, and captions. Write three questions you have about the information you think may be covered in this section.*

1. _____

2. _____

3. _____

Review Vocabulary

subduction

Use your text to define the following term.

_____

_____

New Vocabulary

phytoplankton

amniotic egg

iridium

Use your text to define the following term.

_____

_____

_____

Show your understanding of the term by using it in a sentence of your own.

_____

_____

Academic Vocabulary

define

Define the following term.

_____

⟨**Main Idea**⟩ ———— ⟨**Details**⟩ ————————————————

The Breakup of Pangaea

Use with page 655.

Complete *the flow chart about the breakup of Pangaea.*

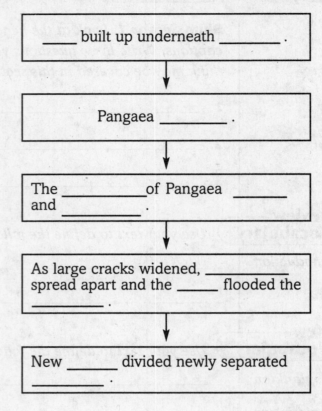

_____ built up underneath _____ .

↓

Pangaea _____ .

↓

The _____ of Pangaea _____ and _____ .

↓

As large cracks widened, _____ spread apart and the _____ flooded the _____ .

↓

New _____ divided newly separated _____ .

Active Tectonism in Western North America

Use with page 657.

Sketch *an outline of North America. Identify which coast had a passive margin and which coast had an active margin during the Mesozoic Era.*

Section 23.2 The Mesozoic Era (continued)

Main Idea —————— **Details** ————————————————————

Active Tectonism in Western North America

Use with page 657.

Organize *information about the three major episodes of orogenies along the western margin of North America during the Mesozoic Era.*

Orogeny	Characteristics of the Orogeny	Area of North America Affected	Era the Orogeny Took Place
Oldest Orogeny			

Mass Extinctions

Use with page 659.

Summarize *the three main pieces of evidence that indicate a meteorite impact caused the Cretaceous-Paleogene mass extinction event.*

1. _____

_____

2. _____

_____

3. _____

_____

SYNTHESIZE New Madrid, Missouri is the site of a failed rift near the center of North America. Hypothesize what would have happened if the rift had not failed. Use your knowledge of rift systems to help you answer this question.

_____

_____

_____

The Paleozoic, Mesozoic, and Cenozoic Eras

Section 23.3 The Cenozoic Era

Main Idea ——————

Details ———————————————

Scan *Section 3 of your text. Use the checklist below.*

- Read all section titles.
- Read all bold words.
- Read all charts and graphs.
- Look at all the pictures and read their captions.
- Think about what you already know about oceans.

Write three facts you discovered about Cenozoic paleogeography as you scanned the section.

1. _____

2. _____

3. _____

Review Vocabulary   *Use your text to define the following term.*

San Andreas Fault _____

_____

New Vocabulary   *Use your text to define each term.*

Homo sapiens _____

_____

bipedal _____

_____

Academic Vocabulary   *Define the following term.*

coincide _____

_____

Section 23.3 The Cenozoic Era (continued)

⟨**Main Idea**⟩———— ⟨**Details**⟩————————————————

Tectonic Events

Use with page 662.

Contrast *the different mountain ranges that were formed and the causes of their formation during the Cenozoic Era. Part of the chart has been completed for you.*

Mountains or Fault Formed	How Formed	When Formed
	subduction of an oceanic plate beneath the western coast of North America	end of _____
San Andreas Fault		
Yellowstone National Park		
		after the breakup of Pangaea

┌─────────────┐
│ **ANALOGY** │ Describe how hands slowly squeezing air out of a ball or balloon
└─────────────┘
can be used as an analogy for the drying up of the Tethys Sea.

_____

_____

_____

_____

_____

_____

The Paleozoic, Mesozoic, and Cenozoic Eras Chapter Wrap-Up

In the "What I Wanted to Find Out" column, copy the questions you listed in the Chapter Preview. In the "What I Learned" column, write down the answers you discovered as you worked through the chapter.

W What I Wanted to Find Out	L What I Learned
1. _____ _____	1. _____ _____
2. _____ _____	2. _____ _____
3. _____ _____	3. _____ _____

Review

Use this checklist to help you study.

☐ Study your Science Notebook for on this chapter.

☐ Study the definitions of vocabulary words.

☐ Review daily homework assignments.

☐ Reread the chapter and review the tables, graphs, and illustrations.

☐ Review the Section Assessment questions at the end of each section.

☐ Look over the Study Guide at the end of the chapter.

SUMMARIZE After reading this chapter, describe three things you have learned about the Paleozoic, Mesozoic, and Cenozoic Eras.

_____

_____

_____

Earth Resources
Chapter Preview

Before you read the chapter, use the "What I Know" column to list three things you know about Earth resources. Then list three questions you have about Earth resources in the "What I Want to Find Out" column.

K What I Know	W What I Want to Find Out
1. _____ _____	1. _____ _____
2. _____ _____	2. _____ _____
3. _____ _____	3. _____ _____

Science Journal

Think of the Launch Lab you did to identify the origins of some common materials and record your responses in this science journal.

Which of the items were identified as coming from living things? Which came from nonliving things?

_____

_____

Which are easily replaced? Which are not replaceable?

_____

_____

What criteria did you use to classify each item? Were you unable to classify any items using just these categories? Explain.

_____

Earth Resources

Section 24.1 Natural Resources

Main Idea ——————— **Details**

Skim *Section 1. Look at the headings, photos, illustrations, and captions. Write three questions you have about the information that you think may be covered in this section.*

1. _____

2. _____

3. _____

Review Vocabulary *Use your text to define the following term.*

water cycle _____

_____

New Vocabulary *Use your text to define each term.*

natural resource _____

_____

renewable resource _____

_____

sustainable yield _____

_____

nonrenewable resource _____

_____

_____

Academic Vocabulary *Define the following term.*

transform _____

Section 24.1 Natural Resources (continued)

⟨**Main Idea**⟩ ————— ⟨**Details**⟩ ———————————————————

Natural Resources

Use with page 678.

Organize *information about natural resources using the information chart.*

Natural Resources

Definition:	What are some examples?
How are they changed?	How can they be classified?

Renewable Resources and Nonrenewable Resources

Use with pages 678–680.

Describe *renewable resources and nonrenewable resources in your own words. Distinguish between the two, and give an example of each one.*

_____

_____

_____

_____

_____

_____

_____

Identify *a renewable resource that exists in an inexhaustible supply.*

_____

_____

Section 24.1 Natural Resources (continued)

| Main Idea | Details |

Renewable Resources and Nonrenewable Resources

Use with pages 678–680.

Classify *each of the following resources as renewable or nonrenewable.*

- air
- carbon
- copper
- diamonds
- fertile soil
- gemstones
- fossil fuels
- gold
- natural gas
- nitrogen
- all living things
- phosphorus
- silver
- solar energy
- trees
- water

Renewable Resources	Nonrenewable Resources

Distribution of Resources

Use with page 680.

Analyze *the graph in Figure 24.4. Write a short paragraph comparing oil use per day in the United States to the rest of the world.*

_____

_____

_____

_____

_____

SYNTHESIZE As fossil fuels like oil become depleted, what do you think will happen to the price of the gasoline we use to power our cars? Explain your reasoning.

_____

_____

_____

Name_____ Date _____

Earth Resources
Section 24.2 Resources from Earth's crust

Main Idea ———— **Details** —————————————————————————

Read *the title of Section 2. List three things that might be discussed in this section.*

1. _____

2. _____

3. _____

Review Vocabulary

Use your text to define the following term.

igneous rock

_____

_____

New Vocabulary

Read the definitions below, then write the term on the blank in the left column.

_____ *naturally accumulating mixture of sand, gravel, and crushed stone found in floodplains, alluvial fans, or glacial deposits*

_____ *mineral that contains a valuable substance that can be mined at a profit;*

_____ *material left after ore is extracted that may release harmful chemicals into surface water or groundwater*

_____ *process by which productive land becomes desert; in arid areas this can occur through the loss of topsoil*

_____ *unweathered, solid parent rock that may consist of limestone, marble, granite, or other quarried rock*

Academic Vocabulary

Define the following term.

volume

_____

Section 24.2 Resources from Earth's Crust (continued)

Main Idea ———— **Details** ————————————

Protected Land

Use with pages 682–683.

Complete *the sentences that discuss land as a natural resource.*

Materials that are derived from land include _____, _____, _____, _____, and _____ . Land provides places for humans and other _____ to _____ .

Much of the land in the United States is certified _____ . These lands exist to protect _____, _____, _____, and _____ .

National forests are managed for _____ . They include areas where _____ are used for many purposes. The national park system protects wildlife _____ and _____ . The national _____ refuges protect habitats and _____ areas for wildlife. Some also provide protection for _____ species.

Bedrock and Aggregates

Use with page 684.

Organize *information about bedrock and aggregates in the chart.*

	Bedrock	Aggregate
What is it made of?		
How is it used?		

Section 24.2 Resources from Earth's Crust (continued)

Main Idea ——— **Details**

Ores

Use with page 684.

Complete *the concept map by listing five ways that ores can form.*

How Ores Form

Other Land Resources

Use with page 685.

Identify *other land resources found on Earth. Describe one use for each resource.*

1._____

2._____

3._____

4._____

5._____

Effects of Mining

Use with page 686.

Analyze *the negative impacts of extracting land resources. Choose one of the negative impacts and connect it to how it could negatively impact you in your everyday life.*

_____

_____

_____

SYNTHESIZE Hypothesize how the natural resources of an area influence the type of housing that is constructed. Provide several examples.

_____

_____

_____

Earth Resources
Section 24.3 Air Resources

(Main Idea)——— **(Details)**———————

Scan *Section 3 of your text. Use the checklist below.*

- Read all section titles.
- Read all bold words.
- Read all charts and graphs.
- Look at all the pictures and read their captions.
- Think about what you already know about air resources.

Write three facts you discovered about air resources as you scanned the section.

1. _____

2. _____

3. _____

(Review Vocabulary) *Use your text to define the following term.*

photosynthesis _____

_____

(New Vocabulary) *Use your text to define each term.*

Nitrogen-fixing bacteria _____

_____

pollutant _____

_____

(Academic Vocabulary) *Define the following term.*

transport _____

Section 24.3 Air Resources (continued)

Main Idea ———— **Details** ————————————————

Origin of Oxygen

Use with page 687.

Sequence *the events that led to oxygen becoming part of Earth's atmosphere.*

_____ Earth's atmosphere contained carbon dioxide, nitrogen, and water vapor.

_____ Rains washed carbon dioxide out of the atmosphere and into the oceans.

_____ Early life forms in the seas used carbon dioxide during photosynthesis.

_____ Oxygen levels in the atmosphere built up.

_____ Earth cooled and became more solid.

_____ Early life forms released oxygen and water vapor.

Cycles of Matter

Use with pages 688–689.

Compare *the nitrogen and carbon cycles.*

_____

_____

_____

_____

_____

_____

_____

_____

_____

_____

Natural Air Pollution Sources

Use with page 690.

Classify *sources of air pollution as coming from these natural origins.*

Volcanoes	Fires	Radon

Section 24.3 Air Resources (continued)

⟨Main Idea⟩ —————— ⟨Details⟩ ————————————————————

Sources of Air Pollution

Use with page 690.

Explain *the importance of having clean air.*

_____

_____

Identify *the four things that can happen to pollutants in the troposphere.*

1. _____

2. _____

3. _____

4. _____

Cycles of Matter

Use with page 688.

Explain *the carbon cycle.*
The student's answer should explain how carbon is converted by green plants and algae by photosynthesis. This carbon is released during respiration.

_____

_____

_____

REAL-WORLD CONNECTION

Sketch a picture of one thing you can do to improve air quality in your area.

Earth Resources

Section 24.4 Water Resources

Main Idea	Details
	Read *the objectives for Section 3. Restate each objective as a question.*
	1. _____
	2. _____
	3. _____
	_____
Review Vocabulary *aquifer*	*Use your text to define the following term.* _____ _____
New Vocabulary *hydrogen bond*	*Use your text to define the following terms.* _____ _____
	Use "desalination" in a sentence that shows its meaning. _____
desalination	_____ _____
Academic Vocabulary *displace*	*Define the following term.* _____

Section 24.4 Water Resources (continued)

⟨ **Main Idea** ⟩—————   ⟨ **Details** ⟩————————————

The Importance of Water

Use with pages 693–694.

Describe *four desirable properties of water.*

1. _____

2. _____

   _____

3. _____

4. _____

Draw *hydrogen bonds using water molecules.*
Use Figure 24.16 for help.

_____

_____

_____

Location of Freshwater Resources

Use with page 694.

Compare *the major uses of freshwater in areas of the United States that receive a lot of precipitation to the major uses of freshwater in areas of the United States that receive little precipitation.*

_____

_____

_____

_____

Section 24.4 Water Resources (continued)

Main Idea	Details

Use of Freshwater Resources

Use with page 694.

Infer *from Figure 24.18 which continent likely receives the most rainfall. Explain your reasoning.*

_____

_____

_____

Managing Freshwater Resources

Use with pages 696–697.

Complete *the graphic organizer about managing freshwater resources.*

```
┌─────────────────────────────────────────────┐
│       Managing Freshwater Resources         │
└─────────────────────────────────────────────┘
   ┌──────┬──────┬──────┬──────┬──────┐
 ┌────┐ ┌────┐ ┌────┐ ┌────┐ ┌────┐
 │    │ │    │ │    │ │    │ │    │
 │    │ │    │ │    │ │    │ │    │
 └────┘ └────┘ └────┘ └────┘ └────┘
```

Identify *ways humans can reduce the need for freshwater resources.*

1._____

2._____

3._____

4._____

5._____

6._____

ANALOGY Construct an analogy between the drawdown of groundwater from an aquifer and the drawdown of a glass of water using drinking straws. Explain your reasoning.

_____

_____

_____

Earth Resources Chapter Wrap-Up

In the "What I Wanted to Find Out" column, copy the questions you listed in the Chapter Preview. In the "What I Learned" column, write down the answers you discovered as you worked through the chapter.

W What I Wanted to Find Out	L What I Learned
1. _____ _____	1. _____ _____
2. _____ _____	2. _____ _____
3. _____ _____	3. _____ _____

Review

Use this checklist to help you study.

☐ Study your Science Notebook for this chapter.

☐ Study the definitions of vocabulary words.

☐ Review daily homework assignments.

☐ Reread the chapter and review the tables, graphs, and illustrations.

☐ Review the Section Assessment questions at the end of each section.

☐ Look over the Study Guide at the end of the chapter.

SUMMARIZE After reading this chapter, describe three things you have learned about Earth resources.

_____

_____

_____

Energy Resources
Chapter Preview

Before you read the chapter, use the "What I Know" column to list three things you know about energy resources. Then list three questions you have about energy resources in the "What I Want to Find Out" column.

K What I Know	W What I Want to Find Out
1. _____ _____	1. _____ _____
2. _____ _____	2. _____ _____
3. _____ _____	3. _____ _____

Science Journal

Think about the Launch Lab you did to observe a type of energy transfer that occurs every day and record your response in this science journal.

Trace the energy source used to bring the water to a boil back to its origin. Describe what happened to the energy as it was used to heat and boil the water. In your description, include an explanation of the source of most energy on Earth. Infer where the energy went when the water began to boil.

_____

_____

_____

_____

_____

_____

Energy Resources
Section 25.1 Conventional Energy Resources

(Main Idea) ─────────── **(Details)** ───────────────────────

Scan *Section 1 of your text. Use the checklist below.*

- Read all section titles.
- Read all bold words.
- Read all charts and graphs.
- Look at all the pictures and read their captions.
- Think about what you already know about conventional energy resources.

Write three facts you discovered about conventional energy resources as you scanned the section.

1. _____

2. _____

3. _____

Review Vocabulary *Use your text to define the following term.*

fault _____

New Vocabulary *Use your text to define each term.*

fuel _____

biomass fuel _____

hydrocarbon _____

peat _____

fossil fuel _____

_____

_____

Academic Vocabulary *Define the following term.*

subsequent _____

Section 25.1 Energy Resources (continued)

⊂Main Idea⊃ ———— ⊂Details⊃ ————————————————

Transfer of Solar Energy

Use with page 708.

Identify *the ultimate source of most energy on Earth.*

_____

Complete *the graphic organizer about the flow of solar energy in an ecosystem. In the circles, give an example of that type of organism in an ecosystem.*

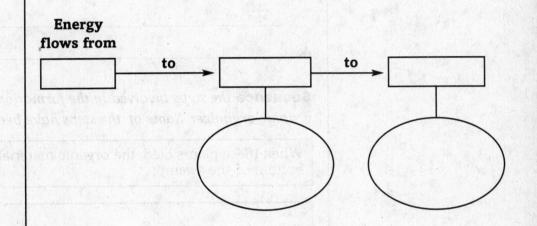

Energy
flows from

Traditional Sources of Energy

Use with page 709.

Identify *traditional fuels. Give an example of each type of fuel whenever possible.*

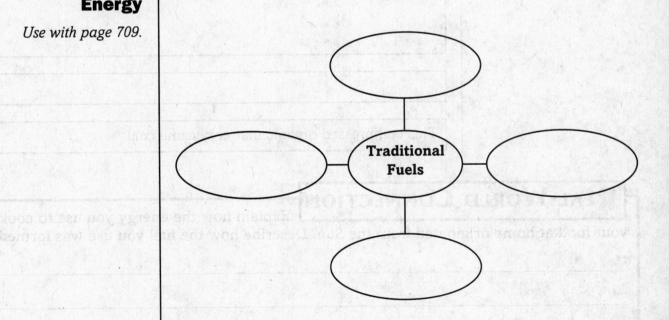

Traditional
Fuels

Section 25.1 Energy Resources (continued)

⸺Main Idea⸺ ———   ⸺Details⸺ ——————————————————

Fossil Fuels

Use with page 711.

List *and describe the four main types of fossil fuels.*

1. _____

2. _____

   _____

3. _____

   _____

4. _____

   _____

Sequence *the steps involved in the formation of coal in the graphic organizer. Some of the steps have been completed for you.*

When these plants died, the organic material settled to the bottom of the swamp.

▼

▼

▼

▼

This compressed organic matter became coal.

REAL-WORLD CONNECTION   Explain how the energy you use to cook your food at home originated from the Sun. Describe how the fuel you use was formed.

_____

_____

_____

Energy Resources
Section 25.2 Alternative Energy Resources

Main Idea	**Details**
	Skim *through Section 2 of your text. Write three questions that come to mind from reading the headings and the illustration captions.*
	1. _____
	2. _____
	3. _____
Review Vocabulary	*Use your text to define the following term.*
electron	_____
New Vocabulary	*Read the definitions below, then write the key term on the blank in the left column.*
_____	*thin, transparent wafers that convert sunlight into electric energy and are made up of layers of boron- and phosphorus-enriched silicon*
_____	*energy generated by converting the energy of free-falling water to electricity.*
	energy produced from Earth's own internal steam and hot water
_____	*the process in which a heavy nucleus divides to form smaller nuclei and one or two neutrons*
Academic Vocabulary	*Define the following term.*
rely	_____

Section 25.2 Alternative Energy Resources (continued)

Main Idea —————— ### Details ———————————————

Solar Energy

Use with page 714.

Compare *how solar energy is collected and used in passive solar heating and in active solar heating.*

_____

_____

_____

_____

Energy from Water

Use with page 716.

Sequence *the steps used in producing hydroelectric power.*

_____ The generator produces electric energy.

_____ The water turns the turbines.

_____ The turbines produce mechanical energy.

_____ Water from a waterfall is diverted into massive turbines.

_____ The mechanical energy drives a generator.

Geothermal Energy

Use with page 717.

Describe *why people are able to bathe in hot springs even though it is cold outside.*

_____

_____

Wind Energy

Use with page 717.

Complete *the flow chart to show how a windmill turns wind energy into usable energy.*

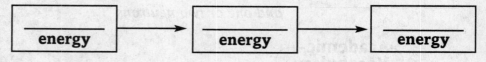

Nuclear Energy

Use with page 718.

Analyze *how the two major nuclear accidents affected the growth of the use of nuclear energy.*

_____

_____

_____

_____

Section 25.2 Alternative Energy Resources (continued)

⟨**Main Idea**⟩────── ⟨**Details**⟩──────────────

Alternative Energy Resources

Use with pages 714–718.

Identify *a major advantage and a major disadvantage of each alternative energy resource.*

Energy Resource	Major Advantage	Major Disadvantage

Biofuels

Use with page 719.

List *and describe the two types of biofuels.*

1._____

_____

2._____

_____

┌─────────────┐
│ **SYNTHESIZE** │ Justify the advantages of one of the alternative energy resources
over its disadvantages.

_____

_____

Energy Resources
Section 25.3 Conservation of Energy Resources

Main Idea	Details
	Read the objectives on the first page of Section 3. Restate each objective as a question that will be answered with information from the section.
	1. _____
	_____
	_____
	2. _____

Review Vocabulary

Use your text to define the following term.

renewable resource

_____

_____

New Vocabulary

Use your text to define each term.

energy efficiency

_____

_____

cogeneration

_____

_____

sustainable energy

_____

_____

_____

Academic Vocabulary

Define the following term.

conclude

_____

Section 25.3 Conservation of Energy Resources (continued)

Main Idea	Details

Energy Efficiency

Use with page 720.

Examine *the graphs in Figure 25.13. Write three questions that can be answered with information from the graphs. Write the answers to each question.*

1. Question: _____

Answer: _____

_____

2. Question: _____

Answer: _____

_____

3. Question: _____

Answer: _____

_____

Use with page 721.

Complete *the concept map about ways energy efficiency can be improved.*

buy _____ , more _____ models

_____ old appliances and vehicles

_____ companies

dependence on _____

Ways to Improve Energy Efficiency

improve efficiency of older _____

add _____
install _____ panels
install new _____
buy new energy _____
efficient _____

government _____

offer _____ savings
fund _____

⎛Main Idea⎞ ——————— ⎛Details⎞ ————————————————————

Getting More for Less

Use with pages 721–723.

Classify *information about improving efficiency in transportation, industry, and at home. List as many things as you can under each heading that can be done to conserve energy.*

Transportation	
Industry	
At Home	

Sustainable Energy

Use with page 723.

Restate *the four key parts of a good plan for global management of Earth's natural resources to meet current and future energy needs without causing environmental damage.*

1. _____   3. _____

2. _____   4. _____

┌─────────────┐
│ **ANALOGY** │ Develop an analogy to explain why it is important to conserve energy
└─────────────┘ resources.

_____

_____

_____

Tie-It-All-Together

FURTHER INQUIRY

Predict *what the main source of fuel in the world will be 100 years from now. Explain your reasoning.*

_____

_____

_____

_____

_____

Connect *If you had to choose an alternative energy resource to develop into an energy resource to be used by the whole United States, which would you choose? Explain your reasoning.*

_____

_____

_____

_____

_____

Develop *a model to collect solar energy near your home. Draw a sketch of your idea and explain how it works.*

_____

_____

_____

Energy Resources Chapter Wrap-Up

In the "What I Wanted to Find Out" column, copy the questions you listed in the Chapter Preview. In the "What I Learned" column, write down the answers you discovered as you worked through the chapter.

W What I Wanted to Find Out	L What I Learned
1._____ _____	1._____ _____
2._____ _____	2._____ _____
3._____ _____	3._____ _____

Review *Use this checklist to help you study.*

☐ Study your Science Notebook for this chapter.

☐ Study the definitions of vocabulary words.

☐ Review daily homework assignments.

☐ Reread the chapter and review the tables, graphs, and illustrations.

☐ Review the Section Assessment questions at the end of each section.

☐ Look over the Study Guide at the end of the chapter.

SUMMARIZE
After reading this chapter, describe three things you have learned about energy resources.

_____

_____

_____

Human Impact on Earth Resources
Chapter Preview

Before you read the chapter, use the "What I Know" column to list three things you know about human impact on the Earth's resources. Then list three questions you have about it in the "What I Want to Find Out" column.

K What I Know	W What I Want to Find Out
1. _____ _____	1. _____ _____
2. _____ _____	2. _____ _____
3. _____ _____	3. _____ _____

Science Journal

Think of the Launch Lab you did to explore the types of resources that you use and some of the global impacts of using them and record your response in this science journal.

How many different resources are represented by the items in your collection?

_____

What are the percentages of renewable and nonrenewable resources?

_____

_____

Where were each of the objects made?

_____

_____

Human Impact on Earth Resources
Section 26.1 Populations and the Use of Natural Resources

Main Idea ———————— **Details** ——————————————————

Skim *Section 1 of your text. Look at the headings, photos, illustrations, and captions. Write three questions you have about the information you think may be covered in this section.*

1. _____

2. _____

3. _____

Review Vocabulary — *Use your text to define the following term.*

population _____

_____

New Vocabulary — *Use your text to define each term.*

exponential growth _____

_____

carrying capacity _____

density–independent factor _____

_____

density–dependent factor _____

_____

Academic Vocabulary — *Define the following term.*

initial _____

Section 26.1 Populations and the Use of Natural Resources (continued)

Main Idea	Details

Resources and Organisms

Use with page 734.

Identify *the four main resources that organisms need.*

1. _____

2. _____

3. _____

4. _____

Describe *three ways humans alter their environment.*

_____

_____

_____

_____

Resources and Population Growth

Use with page 735.

Explain *how as a population increases, its demand for natural resources increases as well.*

_____

_____

_____

_____

Draw *a typical population curve for a population that is experiencing exponential growth.*

Section 26.1 Populations and the Use of Natural Resources (continued)

(Main Idea) —— (Details) ——————————

Limits to Population Growth

Use with pages 735–736.

Describe *what happens to the number of births compared to the number of deaths in the following populations.*

A population below carrying capacity	
A population at carrying capacity	
A population that exceeds carrying capacity	

Classify *factors that affect population growth as either density-dependent factors or density-independent. Draw a circle for each factor you add.*

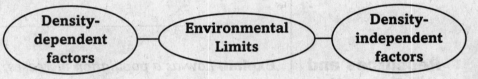

Human Population Growth

Use with page 736.

Identify *the stage that human population growth on Earth is at— early, middle, or late. Explain your reasoning.*

_____

_____

| **SYNTHESIZE** | Describe what would have to happen for the human population on Earth to continue to grow exponentially. Explain why this is impossible. |

_____

_____

_____

Human Impact on Earth Resources
Section 26.2 Human Impact on Earth Resources

⟨Main Idea⟩ ———— **⟨Details⟩** ————————————————————

Scan *Section 2 of your text. Use the checklist below as a guide.*

- Read all section titles.
- Read all bold words.
- Read all tables and graphs.
- Look at all the pictures and read their captions.
- Think about what you already know about human impact on land resources.

Write three facts you discovered about land resources as you scanned the section.

1. _____

2. _____

3. _____

⟨Review Vocabulary⟩ *Use your text to define the following term.*

erosion _____

_____

⟨New Vocabulary⟩ *In the left margin, write the terms defined below.*

_____ *use of organisms to clean up toxic waste*

_____ *chemicals applied to plants to kill insects and weeds*

_____ *process in which a mining company restores land used during mining operations to its original contours and replants vegetation*

_____ *removal of trees from a forested area without adequate replanting, often using clear-cutting, which may result in loss of topsoil and water pollution*

Section 26.2 Human Impact on Earth Resources (continued)

Main Idea ——— **Details** ————————————————————

Extraction of Mineral Resources

Use with pages 737–738.

Explain *the environmental impacts of each type of mineral extraction.*

Surface Mining	→	

Underground Mining	→	

Forestry, Urban Development, Agriculture

Use with pages 739–741.

Organize *the information about the advantages, problems, and solutions associated with agriculture and forestry.*

	Advantages	Problems	Solutions
Forestry			
Crop rotation			
Topsoil			
Pesticides			

Section 26.2 Human Impact on Earth Resources (continued)

Main Idea ————— **Details** ————————————————————————

Urban Development

Use with page 739.

Illustrate *the many ways urban growth impacts the environment. Label each part of your drawing to identify the specific impact shown in that part of your drawing.*

Identify *two methods for cleaning up industrial toxic-waste and briefly describe each.*

1. _____

_____

_____

2. _____

_____

_____

REAL-WORLD CONNECTION

Develop a plan for your family to create less solid waste each day. Explain ways you could change your habits to eliminate some of that solid waste.

_____

_____

Human Impact on Earth Resources
Section 26.3 Human Impact on Air Resources

Main Idea	Details

Consider *the title of Section 3 of your text. List three things that might be discussed in this section.*

1. _____

2. _____

3. _____

Review Vocabulary

Use your text to define the following term.

greenhouse effect

_____

_____

_____

New Vocabulary

Use your text to define each term.

photochemical smog

_____

_____

ozone hole

_____

_____

acid precipitation

_____

_____

_____

Academic Vocabulary

Define the following term.

detect

_____

Section 26.3 Human Impact on Air Resources (continued)

Main Idea

Global Impacts of Air Pollution

Use with pages 743–746.

Details

Compare and Contrast *global warming and the greenhouse effect. Describe how they are similar and different.*

_____

_____

_____

_____

Sequence *the order of reactions that occur as CFCs cause ozone depletion. One of the steps has been completed for you.*

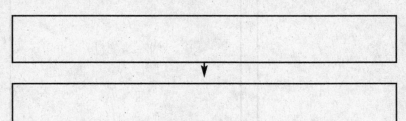

↓

↓

The chlorine atom bonds with an oxygen atom, leaving a molecule of oxygen gas.

↓

↓

↓

Section 26.3 Human Impact on Air Resources (continued)

⟨Main Idea⟩ ———— ⟨Details⟩ ————————————————

Acid Precipitation

Use with page 745.

Organize *the causes and effects of acid precipitation.*

Causes of Acid Precipitation	Effects of Acid Precipitation

Reducing Air Pollution

Use with page 747.

Identify *four ways air pollution has been or can be reduced.*

1. _____

2. _____

3. _____

4. _____

_____

ANALOGY   Formulate an analogy you can use to explain the greenhouse effect to a group of younger students. Think about how the greenhouse effect warms Earth and what kind of experiment could be used to show this.

_____

_____

_____

Human Impact on Earth Resources

Section 26.4 Human Impact on Water Resources

Main Idea	Details
	Preview *Section 4 of your text. Read the headings and captions of photographs and illustrations. Write three questions you think may be answered by the information in this section.*
	1. _____ _____
	2. _____ _____
	3. _____ _____
Review Vocabulary *runoff*	*Use your text to define the following term.* _____ _____ _____
New Vocabulary *point source*	*Use your text to define each term.* _____ _____ _____
nonpoint source	_____ _____ _____
Academic Vocabulary *generate*	*Define the following term.* _____

Section 26.4 Human Impact on Water Resources (continued)

Main Idea ———— **Details** ——————————————————

**Water
Conservation**

Use with page 748.

Identify *ways to conserve water. In each box, list ways water can
be conserved in each of those areas.*

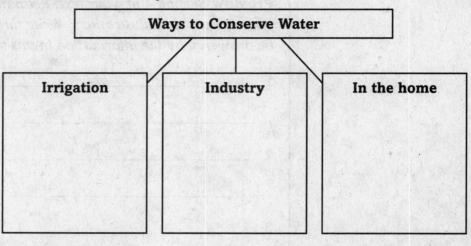

Water Pollution

Use with page 749.

Contrast *point sources of water pollution and nonpoint sources of
water pollution. Give an example of each.*

_____

_____

_____

Describe *the causes and effects of different types of water pollution.*

	Causes	Effects
Pollution of Groundwater		
Pollution in Oceans		

Section 26.4 Human Impact on Water Resources (continued)

Main Idea	**Details**
Reducing Water Pollution *Use with page 750.*	**Describe** *the two laws passed in the United States to fight water pollution. Explain the purpose of each law and describe how each law is working.* 1._____ _____ _____ 2._____ _____ _____ _____ _____ **Analyze** *how your like has been impacted by the Safe Drinking Water Act and the Clean Water Act.* _____ _____ _____ _____ _____

SYNTHESIZE Create a journal entry about how you used water today. Also include an explanation of how you conserved water today.

Today I used water to... _____

_____

Human Impact on Earth Resources   Chapter Wrap-Up

In the "What I Wanted to Find Out" column, copy the questions you listed in the Chapter Preview. In the "What I Learned" column, write down the answers you discovered as you worked through the chapter.

W What I Wanted to Find Out	L What I Learned
1. _____ _____	1. _____ _____
2. _____ _____	2. _____ _____
3. _____ _____	3. _____ _____

Review   *Use this checklist to help you study.*

☐ Study your Science Notebook for this chapter.

☐ Study the definitions of vocabulary words.

☐ Review daily homework assignments.

☐ Reread the chapter and review the tables, graphs, and illustrations.

☐ Review the Section Assessment questions at the end of each section.

☐ Look over the Study Guide at the end of the chapter.

SUMMARIZE
After reading this chapter, describe three things you have learned about how humans impact Earth resources.

_____

_____

_____

The Sun-Earth-Moon System
Chapter Preview

Before you read the chapter, use the "What I Know" column to list three things you know about the Sun-Earth-Moon system. Then list three questions you have about this system in the "What I Want to Find Out" column.

K What I Know	W What I Want to Find Out
1. _____	1. _____
2. _____ _____	2. _____ _____
3. _____ _____	3. _____ _____

Science Journal

Think about the Launch Lab you did to compare relative sizes and distances within the Sun-Earth-Moon system and record your response in this science journal.

Describe the sizes of your cutout Earth and Moon compared to the distance between them.

_____

_____

Infer why you were not instructed to cut out a scaled Sun and place it at the scaled distance.

_____

_____

How would you change this model so that it would fit in your classroom?

_____

_____

The Sun-Earth-Moon System

Section 27.1 The Tools of Astronomy

⊂Main Idea⊃ ———   **⊂Details⊃** ———————————————

Scan *Section 1 of your text. Use the checklist below as a guide.*

• Read all section titles.

• Read all bold words.

• Look at all the pictures and read their captions.

• Think about what you already know about the tools of astronomy.

Write three facts you discovered about the tools that scientists use to observe objects in space.

1. _____

_____

2. _____

3. _____

⊂New Vocabulary⊃   *Use your text to define each term.*

eletromagnetic spectrum _____

_____

refracting telescope _____

_____

reflecting telescope _____

_____

interferometry _____

_____

⊂Academic Vocabulary⊃   *Define the following term.*

adapt _____

_____

Section 27.1 The Tools of Astronomy (continued)

⟨**Main Idea**⟩——— ⟨**Details**⟩————————————————

Radiation

Use with page 764.

Sequence *each type of electromagnetic radiation by order of decreasing wavelength. Then write one fact about each type of electromagnetic radiation in the space to the right.*

_____ Gamma rays: _____

_____ Infrared radiation: _____

_____ Microwaves: _____

_____ Radio waves: _____

_____ Ultraviolet radiation: _____

_____ Visible light: _____

_____ X-rays: _____

_____

Telescopes

Use with page 765.

Identify *the major benefits of telescopes by completing the graphic organizer below.*

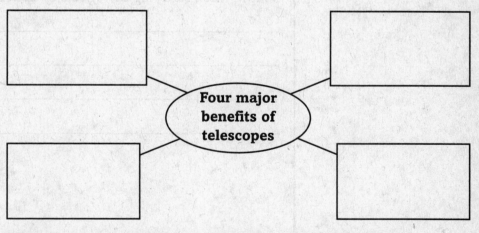

Four major benefits of telescopes

Sketch *how a magnifying glass collects light and brings it to a focus. Refer to Figure 27.3 for help.*

Name _____ Date _____

Section 27.1 The Tools of Astronomy (continued)

Main Idea · Details

Satellites, Probes, and Space-Based Astronomy

Use with pages 767–769.

Classify *Complete the chart by writing **telescope**, **satellite**, or **robot** beside its description. Then give two examples of each.*

Description	Object	Two examples
Observes wavelengths blocked by Earth's atmosphere		
Collects light from a distant object and focuses it at a point.		
Collects objects from other planets		

Describe *one advantage that space stations have over the space shuttle when studying the effects of space.*

_____

_____

_____

_____

REAL-WORLD CONNECTION
Payload is the term astronomers use to describe all the equipment a satellite needs to do its job. Infer from the information in your textbook what kinds of equipment might be on the Hubble Space Telescope.

_____

_____

320 *Section 27.1 The Tools of Astronomy*

The Sun-Earth-Moon System
Section 27.2 The Moon

Main Idea ————— **Details** ——————————————————————

Consider the title of Section 2. List three things that might be discussed in this section.

1. _____

2. _____

3. _____

Review Vocabulary

lava

Use your text to define the following term.

_____

_____

New Vocabulary

In the left margin, write the terms defined below.

_____ 1. the amount of sunlight that the surface of the Moon reflects

_____ 2. mountainous regions of the Moon

_____ 3. dark, smooth plains of the Moon

_____ 4. craters that formed when objects from space crashed onto the Moon's surface

_____ 5. material that blasted out of the Moon's surface and fell back to the surface

_____ 6. long trails of ejecta

_____ 7. Structures that resemble valleys

_____ 8. Loose, ground up rock on the surface of the Moon

Academic Vocabulary

unique

Define the following term.

_____

Section 27.2 The Moon

⟨Main Idea⟩ ————

⟨Details⟩ ————————————————————

Reaching for the Moon

Use with page 770.

Create *a time line of the following developments in Moon exploration. Be sure to mark the year that each development took place.*

• Alan B. Shepard Jr. becomes first American in space.

• The Soviet Union launches *Sputnik I.*

• *Apollo 11* lands on the Moon.

• Yuri A. Gagarin becomes first human in space.

• Plans for a crewed lunar expedition begin.

Use with page 772.

Describe *four unique features of Earth's moon.*

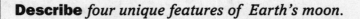

1. _____

2. _____

_____

3. _____

_____

4. _____

Section 27.2 The Moon

<Main Idea> ——   <Details> ——

Use with pages 771–772.

Organize *the characteristics of Earth and the Moon in the Venn diagram.*

• has an average albedo of about 7 percent

• has an average albedo of about 31 percent erosion

• mineral make up is mostly silicates

• surface has not been altered by erosion

• has no flowing water

• has very few craters

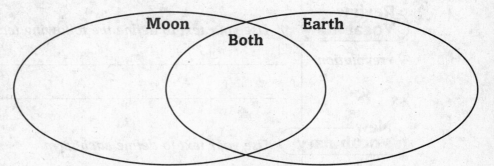

Moon   Both   Earth

History of the Moon

Use with page 772.

Sequence *the stages in the process of the impact theory.*

____ Material merged together to form the Moon.

____ The impact threw material from the body and Earth into space.

____ A mars-sized body collided with Earth.

Point out one piece of evidence that supports the idea of the impact theory.

_____

_____

_____

REAL-WORLD CONNECTION Consider the following statement: The
study of Earth's surface provides us with a more accurate history of our solar system
than the Moon's surface. Explain why you agree or disagree with this statement.

_____

_____

_____

The Sun-Earth-Moon System
Section 27.3 The Sun-Earth-Moon System

(Main Idea)	(Details)

Read *the title of Section 3. List three things that might be discussed in this section.*

1. _____

2. _____

3. _____

Review Vocabulary *Use your text to define the following term.*

revolution
_____
_____

New Vocabulary *Use your text to define each term.*

ecliptic plane _____

solstice _____

equinox _____

_____

synchronous rotation _____

_____

solar eclipse _____

perigee _____

apogee _____

lunar eclipse _____

Academic Vocabulary *Define the following term.*

distribute _____

Section 27.3 The Sun-Earth-Moon System

⌐Main Idea⌐	⌐Details⌐

Daily Motions

Use with page 775.

Complete *the flow chart below to show how a Focault pendulum demonstrates that Earth is rotating.*

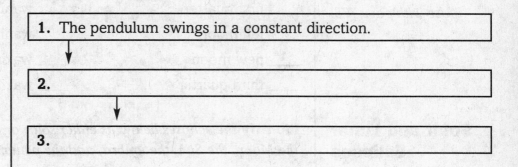

1. The pendulum swings in a constant direction.

2.

3.

Sketch *your own example of the Coriolis effect using air flow. Be sure to label your sketch and use arrows to indicate air direction. Briefly explain your work in the lines below.*

_____

_____

_____

Annual Motions

Use with page 776.

Describe *the tilt of Earth's axis during winter and summer in the northern hemisphere.*

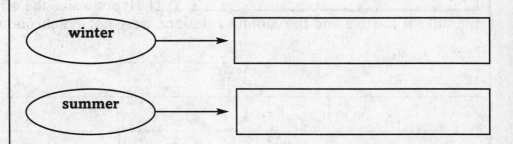

winter ⟶

summer ⟶

Section 27.3 The Sun-Earth-Moon System

Main Idea ———— **Details** ————————————————

Phases of the Moon

Use with page 778.

Sequence *the phases of the moon as it orbits Earth. The first phase has been done for you.*

____ first quarter   ____ waning crescent

____ full moon   ____ waning gibbous

__1__ new moon   ____ waxing crescent ·

____ third quarter   ____ waxing gibbous

Solar and Lunar Eclipses

Use with pages 779–782.

Identify *the eclipses below as either solar or lunar. Then label Earth, the Moon, the Sun, the umbra, and the penumbra in each figure.*

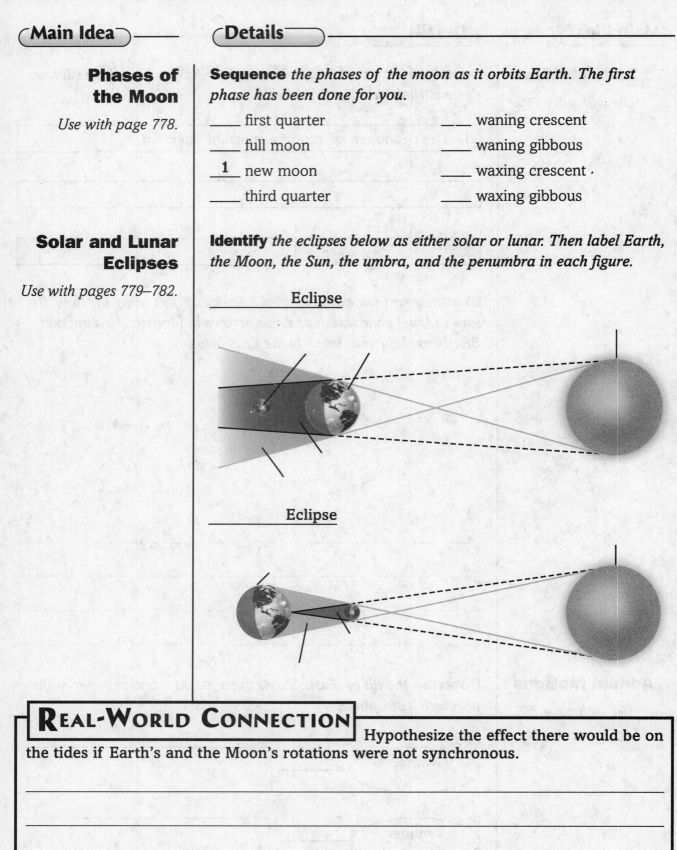

Eclipse

Eclipse

REAL-WORLD CONNECTION Hypothesize the effect there would be on the tides if Earth's and the Moon's rotations were not synchronous.

_____

_____

_____

Tie-It-All-Together

Relate *how you think the phrase "the dark side of the Moon" originated.*

_____

_____

_____

_____

Compare *the Moon's properties with the properties of one other planet in our solar system. How are they different? How are they the same?*

_____

_____

_____

_____

Design *a page for a middle school science book describing the Moon's effect on Earth's tides. Explain in simple terms how the Moon effects the tides. Include a sketch with labels and a caption. Share your page with the class.*

_____

_____

_____

_____

The Sun-Earth-Moon System Chapter Wrap-Up

In the "What I Wanted to Find Out" column, copy the questions you listed in the Chapter Preview. In the "What I Learned" column, write down the answers you discovered as you worked through the chapter.

W What I Wanted to Find Out	L What I Learned
1. _____ _____	1. _____ _____
2. _____ _____	2. _____ _____
3. _____ _____	3. _____ _____

Review *Use this checklist to help you study.*

☐ Study your Science Notebook for this chapter.

☐ Study the definitions of vocabulary words.

☐ Review daily homework assignments.

☐ Reread the chapter and review the tables, graphs, and illustrations.

☐ Review the Section Assessment questions at the end of each section.

☐ Look over the Study Guide at the end of the chapter.

SUMMARIZE After reading this chapter, list three things you have learned about Earth, the Sun, and the Moon.

_____

_____

_____

Our Solar System
Chapter Preview

Before you read the chapter, use the "What I Know" column to list three things you know about fossils. Then list three questions you have about fossils in the "What I Want to Find Out" column.

K What I Know	W What I Want to Find Out
1. _____ _____	1. _____ _____
2. _____ _____	2. _____ _____
3. _____ _____	3. _____ _____

Science Journal

Think about the Launch Lab you did to learn about space missions and their discoveries and record your responses in this science journal.

Make an outline for each mission. Include the type of mission (flyby, lander, or orbiter), the scientific goals, the launch date and the date of arrival at the planet, and a summary of what was learned, or what scientists hope will be learned.

_____

_____

_____

_____

_____

Our Solar System

Section 28.1 Formation of the Solar System

(Main Idea)	(Details)

Skim *Section 1 of your text. Focus on the headings and the illustration captions. List three topics that might be discussed in this section.*

1. _____

2. _____

3. _____

Review Vocabulary   *Use your text to define the following term.*

focus   _____

New Vocabulary   *Use your text to define each term.*

planetisimal   _____

retrograde motion   _____

ellipse   _____

_____

astronomical unit   _____

_____

eccentricity   _____

_____

Academic Vocabulary   *Define the following term.*

focus   _____

_____

Section 28.1 Formation of the Solar System (continued)

⊂Main Idea⊃————

A Collapsing Interstellar Cloud

Use with pages 796–797.

⊂Details⊃————————————

Sequence *the events of a collapsing interstellar cloud. The first one has been done for you.*

___ The cloud becomes denser at the center.

___ Rotation slows and the cloud flattens

___ The cloud spins faster and faster.

___ The cloud becomes a rotating disk.

_1_ The collapse of the cloud begins to accelerate.

Compare *physical data of the planets. Reference Table 28.1 on page 798.*

Planet	Diameter (km)	Atmosphere	Distance from the Sun (AU)
Mercury	4,880		
Venus			0.72
Earth			
Mars		CO_2, A_r, N_2, O_2	
Jupiter			
Saturn			9.58
Uranus	50,724		
Neptune			30.4

Analyze *the difference in planet atmosphere in relation to distance from the Sun. Explain the pattern you see.*

_____

_____

_____

_____

_____

Section 28.1 Formation of the Solar System (continued)

Main Idea ———— **Details** ——————————————————

Gravity and Orbits

Use with pages 802–803.

Explain *Newton's law of universal gravitation by completing the following sentence.*

Every pair of bodies in the universe _____ each other

with a _____ that is proportional to the _____ of

their masses and inversely _____ to the square of the

distance between them.

Illustrate *the center of mass using two people on a see-saw, where one person is much heavier than the other. Be sure to include the balance point in your sketch. Refer to Figure 28.9 on page 803 for help.*

Compare *the masses of two people on a see-saw if the balance point is near the middle of the see-saw. Explain your reasoning.*

_____

_____

ANALOGY Relate how one car passing another car on a highway might be similar to the retrograde motion of planets.

_____

_____

_____

_____

Our Solar System
Section 28.2 The Inner Planets

(**Main Idea**)——— (**Details**)——————————————

Scan *Section 2 of your text. Use the checklist below as a guide.*

- Read all section titles.
- Read all bold words.
- Read all tables.
- Look at all the photos of each planet and read their captions.
- Think about what you already know about Mercury, Venus, Mars, and Earth.

Write three facts you discovered about Mercury, Venus, Mars, and Earth as you scanned the section.

1. _____

2. _____

3. _____

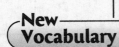
Review Vocabulary   *Use your text to define the following term.*

albedo   _____

_____

New Vocabulary   *Use your text to define the each term.*

terrestrial planets   _____

scarp   _____

Academic Vocabulary   *Define the following term.*

similar   _____

| Main Idea | Details |

Mercury, Venus, Earth, Mars

Use with pages 804–810.

Describe *four characteristics of Venus using the graphic organizer.*

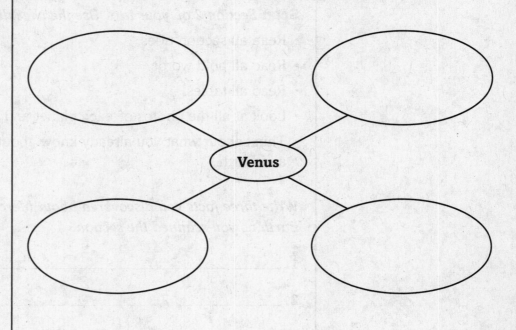

Compare *the features of the terrestrial planets by completing the table below.*

Feature	Mercury	Venus	Earth	Mars
Relative position from the Sun				
Size				
Atmosphere				
Weather				
Surface				
# of moons				

Section 28.2 The Inner Planets (continued)

Main Idea ————

Mercury, Venus, Earth, and Mars

Use with pages 804–810.

Details ——————————————————————

Describe *Mercury by completing the following sentences.*

Unlike Earth, Mercury has little _____ . What does exist is

made up of mostly _____ and _____ . Mercury's surface is similar

to the _____ surface because it is covered with craters and plains.

The surface gravity is _____ than that of the Moon. The

_____ of Mercury suggests the planet has a large

_____ core.

Identify *three reasons why life can exist on Earth.*

1. _____

2. _____

3. _____

Create *a postcard describing the atmosphere and the surface of Mars. Imaging that you are visiting Mars and sending your postcard to a friend at home.*

REAL-WORLD CONNECTION Hypothesize what conditions on the terrestrial planets would be like today if their atmospheres had been warmer and richer in carbon dioxide.

_____

_____

_____

Our Solar System
Section 28.3 The Outer Planets

Main Idea	Details
	Consider *the title of Section 3. List three topics that might be discussed in this section.*
	1. _____
	2. _____
	3. _____
Review Vocabulary	*Use your text to define the following term.*
asteroid	_____
	_____
New Vocabulary	*Use your text to define the each term.*
gas giant planet	_____
liquid metallic hydrogen	_____
	_____
belt	_____
	_____
zone	_____
	_____
Academic Vocabulary	*Define the following term.*
hypothesize	_____
	_____
	_____

Section 28.3 The Outer Planets (continued)

Main Idea —— **Details** ————————————————————

Jupiter

Use with pages 811–812.

Compare *the four largest moons of Jupiter by completing the table.*
State at least three facts about each moon.

Io	Europa	Ganymede	Callisto

Saturn

Use with page 813.

Describe *Saturn's rings by completing the following paragraph.*

Saturn's rings are composed of pieces of _____ and _____

that range from _____ to the size of _____ . There

are _____ major rings, but each ring is actually made up of

narrower rings, called _____ , and many open _____ .

Uranus and Neptune

Use with pages 814–815.

Compare *the characteristics of Uranus and Neptune*

_____

_____

_____

_____

_____

_____

Section 28.3 The Outer Planets (continued)

⟨Main Idea⟩ ——————   ⟨Details⟩ ——————————————————

Use with pages 814–815.

Describe *several characteristics of each planets atmosphere and moons and rings.*

	Uranus	Neptune
Atmosphere		
Moons and Rings		

REAL-WORLD CONNECTION

There is a possibility that Europa, one of Jupiter's moons, has oceans. Astronomers already know that Europa has a variety of basic chemicals and a source of heat energy. Assess the possibility that life may have developed on Europa. Explain your reasoning.

_____

_____

Our Solar System
Section 28.4 Other Solar System Objects

◖Main Idea◗ ——— **◖Details◗** ——————————————————————

Scan *Section 4 of your text. Use the checklist below to preview.*

- Read all section titles.
- Read all bold words.
- Read all tables and graphs.
- Look at all the pictures and read their captions.
- Think about what you already know about how the formation of our solar system.

Write three facts you learned about the formation of our solar system as you scanned the section.

1. _____

2. _____

3. _____

◖New Vocabulary◗   *In the left margin, write the terms defined below.*

_____   *an object that, due to its own gravity, is spherical in shape and orbits the sun, but is not a satellite*

_____   *interplanetary material that falls toward Earth and enters Earth's atmosphere*

_____   *streak of light produced as a meteoroid burns up in Earth's atmosphere*

_____   *meteoroid that strikes Earth's surface*

_____   *rocky asteroidlike belt that lies outside the orbit of Neptune*

_____   *small, icy bodies that have eccentric orbits around the Sun*

_____   *occurs when comet particles burn up entering Earth's upper atmosphere*

◖Academic Vocabulary◗   *Define the following term.*

eventual   _____

Section 28.4 Other Solar System Objects (continued)

⊂**Main Idea**⊃ ——— ⊂**Details**⊃ ————————————————

Pluto

Use with page 816.

Identify *five unique characteristics of Pluto.*

1. _____

2. _____

3. _____

4. _____

5. _____

Kuiper Belt Objects

Use with page 817.

Compare *the characteristics of Kuiper Belt objects in the data table below. Refer to Figure 28-27 for help.*

Object	Total period years	Distance from Sun (AU)	Color
	240	about 30	
Sedna	10,500		
		about 97	white
El 61	285		
Buffy			

Section 28.4 Other Solar System Objects (continued)

Main Idea — **Details**

Asteroids

Use with page 818.

Identify *the events that are associated with the formation of a meteorite by filling in the blanks below.*

_____ orbit, collide, and break into fragments.

⬇

These fragments enter Earth's atmosphere and become _____ .

⬇

Meteoroids burn up in Earth's atmosphere, producing a streak of light called a _____ .

⬇

If the meteoroid does not completely burn up, part of it will collide with the ground and is then called a _____ .

Comets

Use with page 819.

Sketch *a comet in the space below. Label the nucleus, coma, and tail.*

REAL-WORLD CONNECTION

Draw a time line starting 4.6 billion years ago. Indicate the times of the formation of Earth and the Moon, the appearance of oceans on Earth, and the beginnings of life on Earth. Explain what you notice about the location of these events.

Our Solar System Chapter Wrap-Up

In the "What I Wanted to Find Out" column, copy the questions you listed in the Chapter Preview. In the "What I Learned" column, write down the answers you discovered as you worked through the chapter.

W What I Wanted to Find Out	L What I Learned
1. _____ _____	1. _____ _____
2. _____ _____	2. _____ _____
3. _____ _____	3. _____ _____

Review

Use this checklist to help you study.

☐ Study your Science Notebook for this chapter.

☐ Study the definitions of vocabulary words.

☐ Review daily homework assignments.

☐ Reread the chapter and review the tables, graphs, and photos.

☐ Review the Section Assessment questions at the end of each section.

☐ Look over the Study Guide at the end of the chapter.

SUMMARIZE After reading this chapter, list three things you have learned about our solor system.

_____

_____

_____

Stars

Chapter Preview

Before you read the chapter, use the "What I Know" column to list three things you know about stars. Then list three questions you have about stars in the "What I Want to Find Out" column.

K What I Know	W What I Want to Find Out
1. _____ _____	1. _____ _____
2. _____ _____	2. _____ _____
3. _____	3. _____

Science Journal

Think of the Launch Lab you did to monitor the Sun and record your response in this science journal.

Is the sun near the peak of its activity cycle?

_____

_____

_____

Do the solar images that you observed fit with your expectation of where the Sun is in its activity cycle? Compare and contrast the images obtained at different wavelengths.

_____

_____

_____

_____

Stars

Section 29.1 The Sun

⊂Main Idea⊃	⊂Details⊃

Skim *Section 1 of your text. Write three questions that come to mind from reading the headings and the illustration captions.*

1. _____

2. _____

3. _____

Review Vocabulary

Use your text to define the following term.

magnetic field _____

_____

New Vocabulary

In the left margin, write the terms defined below.

_____ the lowest visible layer of the Sun's atmosphere and the surface of the Sun

_____ layer of the Sun's atmosphere above the photosphere and below the corona

_____ top layers of the Sun's atmosphere

_____ a stream of high-speed, ionized particles flowing outward through the solar system from the Sun's corona

_____ relatively cool, dark spots appearing on the surface of the photosphere

_____ a sudden eruption of particles and radiation from the surface of the Sun

_____ an arc of flaming gas rising from the chromosphere

_____ a nuclear reaction in which lightweight nuclei combine into heavier nuclei

_____ a nuclear reaction in which an atomic nucleus splits into smaller, lighter nuclei releasing energy

Section 29.1 The Sun (continued)

Properties of the Sun

Use with page 830.

Identify *four unique characteristics of the Sun.*

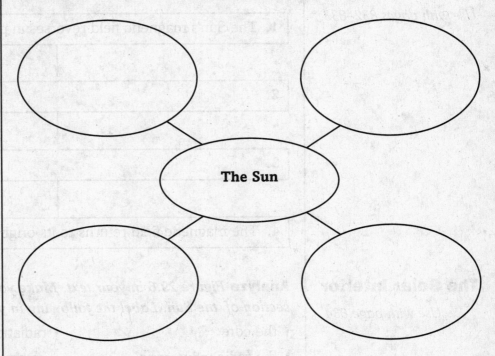

The Sun

The Sun's Atmosphere

Use with pages 831–832.

State *three facts about each layer of the Sun's atmosphere in the table below.*

Photosphere	Chromosphere	Corona
1.	1.	1.
2.	2.	2.
3.	3.	3.

Section 29.1 The Sun (continued)

◖**Main Idea**◗ ——— ◖**Details**◗ ————————————————

Solar Activity

Use with pages 832–833.

Sequence *the solar activity cycle by completing the flow chart.*

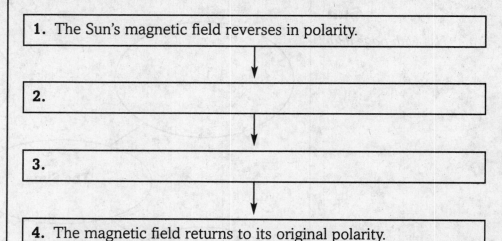

1. The Sun's magnetic field reverses in polarity.

↓

2.

↓

3.

↓

4. The magnetic field returns to its original polarity.

The Solar Interior

Use with page 834.

Analyze *Figure 29.6 in your text. Make your own sketch of a cross section of the Sun. Label the following in your sketch:*

- the core
- the radioactive zone
- the convective zone

- radiation
- convection

Section 29.1 The Sun (continued)

Main Idea	Details

Spectra

Use with page 835.

Compare *continuous, emission, and absorption spectra. Write each characteristic in the correct location in the diagram below.*

- appears as bright lines
- has no breaks in spectrum
- appears as a series of dark bands

- produced by a solid, liquid, or gas
- comes from a non-compressed gas
- visible light arranged by wavelength

Emission   Continuous

All

Absorbtion

Solar Composition

Use with page 836.

Explain *solar composition by completing the sentence.*

The Sun consists of _____ , about ____ percent by mass,

and _____ , ____ percent, as well as a small amount of other

_____ .

REAL-WORLD CONNECTION

Hypothesize *how the planets would be affected if the Sun did not make up more than 99% of the mass of the entire solar system.*

_____

_____

_____

Stars

Section 29.2 Measuring the Stars

Main Idea	Details
	Consider the objectives on the first page of the section. Write three questions you think will be answered in the section based on the objectives.

1. _____

2. _____

3. _____

Review Vocabulary Use your text to define the following term.

wavelength _____

New Vocabulary Use your text to define each term.

constellation _____

binary star _____

_____

parallax _____

parsec _____

apparent magnitude _____

_____

absolute magnitude _____

luminosity _____

Hertzsprung-Russell diagram _____
main sequence _____

_____

Academic Vocabulary Define the following term.

isolate _____

Section 29.2 Measuring the Stars (continued)

Main Idea ——— **Details** ————————————

Groups of Stars

Use with pages 837–841.

Organize *information about star clusters by completing the diagram below.*

Star clusters are

either / or

_____ clusters   _____ clusters

because the stars   because the stars

are not _____ packed   are densely packed into a _____ shape

Basic Properties of Stars

Use with pages 841–842.

Identify *the six basic properties that scientists use to categorize stars.*

1._____ 4._____

2._____ 5._____

3._____ 6._____

Sequence *the celestial objects in order of increasing absolute magnitude. The first one has been done for you.*

____ Sirius   ____ most luminous galaxies

____ most luminous stars   ____ Uranus

_1_ Pluto   ____ Venus

____ Sun   ____ Full moon

Section 29.2 Measuring the Stars (continued)

(Main Idea)	(Details)

Spectra of Stars

Use with page 844.

Write *review question related to the classification of stars. Then write a second review question related to wavelength shifts. Give the answer to each of your questions.*

1. Question: _____

Answer: _____

_____

2. Question: _____

Answer: _____

_____

Analyze *Table 29-2 in your text. Describe the general characteristics of stars found in different locations on the H-R diagram. The first one has been done for you.*

Upper-left corner	Upper right-corner	Lower-left corner
large, hot and luminous		

SUMMARIZE *In your own words, summarize how scientists calculate distances from* Earth to stars.

_____

_____

_____

_____

_____

_____

Stars
Section 29.3 Stellar Evolution

Main Idea ———— **Details** ————————————————

Scan *Section 3 of your text. Use the checklist below to preview.*

- Read all section titles.
- Read all bold words.
- Read all tables and graphs.
- Look at all the pictures and read their captions.
- Think about what you already know about stellar evolution.

Write three facts you discovered about stellar evolution as you scanned the section.

1. _____

2. _____

3. _____

Review Vocabulary — *In the left margin, write the terms defined below.*

_____ *a cloud of interstellar dust or gas or both, which collapses on itself as a result of its own gravity*

_____ *the hot condensed disk-shaped object at the center of a collapsing cloud*

_____ *collapsed, dense core of a star that forms quickly while its outer layers are falling inward*

_____ *a spinning nuetron star that exhibits a pulsing pattern*

_____ *massive explosion that occurs when the outer layers of a star are blown off*

_____ *small, extremely dense remnant of a star whose gravity is so immense that not even light can escape its gravity field*

Academic Vocabulary — *Define the following term.*

dynamic | _____

Section 29.3 Stellar Evolution (continued)

⟨Main Idea⟩ ——— **⟨Details⟩** —————————————————————————————————

Basic Structure of Stars

Use with page 847.

Imagine *you were explaining hydrostatic equilibrium to a middle school student. Draw a diagram in the space below that you could use to illustrate the concept, then write a brief explanation.*

_____

_____

_____

_____

_____

Stellar Evolution and Life Cycles

Use with pages 847–848.

Sequence *the steps in the formation of a star by completing the flow chart.*

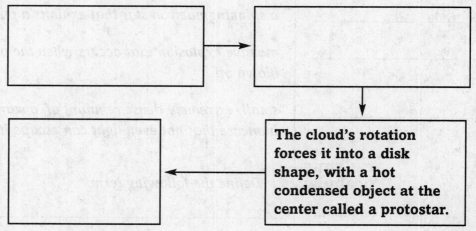

The cloud's rotation forces it into a disk shape, with a hot condensed object at the center called a protostar.

Section 29.3 Stellar Evolution (continued)

Main Idea — **Details**

The Sun's Life Cycle

Use with pages 848–849.

Sequence *the steps in the life cycle of the Sun.*

_____ converts hydrogen into helium in its core

_____ star contracts back to normal size and becomes stable

_____ As the core runs out of hydrogen and then helium, the core contacts and the outer layers expand, cool, and become less bright.

_____ Stars are born in nebulae.

_____ outer layers expand again and are driven off entirely leaving a planetary nebula

_____ the star becomes a red giant, the core becomes hot enough for the core to become carbon

_____ the core becomes exposed and is a white dwarf

_____ the star loses gas from its outer layers

_____ Huge clouds of dust and gas collapse under gravitational forces, forming protostars.

Life Cycles of Massive Stars

Use with pages 849–851.

Compare *the evolution of a massive star with the evolution of the Sun. Write each description in the correct place in the Venn diagram.*

- begins as a nebula
- becomes a red giant once
- becomes a red giant several times
- fewer reaction phases
- loses less of its mass
- loses much of its mass
- more reaction phases

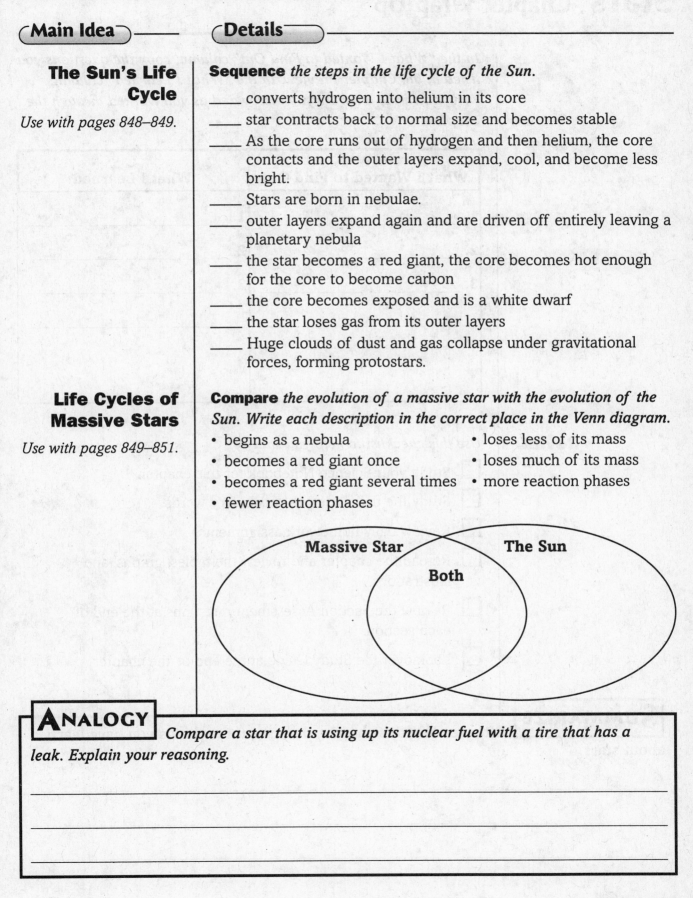

Massive Star   Both   The Sun

ANALOGY *Compare a star that is using up its nuclear fuel with a tire that has a leak. Explain your reasoning.*

_____

_____

_____

Stars Chapter Wrap-Up

In the "What I Wanted to Find Out" column, copy the questions you listed in the Chapter Preview. In the "What I Learned" column, write down the answers you discovered as you worked through the chapter.

W What I Wanted to Find Out	L What I Learned
1. _____	1. _____
2. _____	2. _____
3. _____	3. _____
_____	_____

Review   *Use this checklist to help you study.*

☐ Study your Science Notebook for this chapter.

☐ Study the definitions of vocabulary words.

☐ Review daily homework assignments.

☐ Reread the chapter and review the tables, graphs, and illustrations.

☐ Review the Section Assessment questions at the end of each section.

☐ Look over the Study Guide at the end of the chapter.

SUMMARIZE   After reading this chapter, describe three things you have learned about stars.

_____

_____

_____

Galaxies and the Universe
Chapter Preview

Before you read the chapter, use the "What I Know" column to list three things you know about galaxies and the universe. Then list three questions you have about them in the "What I Want to Find Out" column.

K What I know	W What I Want to Find Out
1. _____ _____	1. _____ _____
2. _____ _____	2. _____ _____
3. _____ _____	3. _____ _____

Science Journal

Think about the Launch Lab you did to model the Milky Way.

Describe what your model of the Milky Way would look like if you actually built it.

_____

_____

Explain why it would be a problem to show the size of our solar system in comparison to the Milky Way.

_____

_____

Explain how you would change your model to include the size of Earth.

_____

_____

Galaxies and the Universe

Section 30.1 The Milky Way Galaxy

Main Idea	Details
	Scan *Section 1 of your text. Use the checklist below as a guide.*

- Read all section titles.
- Read all bold words.
- Read all graphs and equations.
- Look at all the pictures and read their captions.

Write three facts you discovered about the Milky Way as you scanned the section.

1. _____

2. _____

3. _____

Review Vocabulary

Use your text to define the following term.

galaxy _____

_____

New Vocabulary

Use your text to define each term.

variable star _____

RR Lyrae variable _____

Cepheid variable _____

halo _____

Population I stars _____

Population II stars _____

spiral density wave _____

Academic Vocabulary

Define the following term.

interpret _____

Section 30.1 The Milky Way Galaxy (continued)

Main Idea	Details
Discovering the Milky Way, The Shape of the Milky Way *Use with pages 860–862.*	**Sketch** *the Milky Way below. Be sure to include the following parts:* • Sun • Halo • Globular clusters • Disk • Nuclear bulge

Mass of the Milky Way

Use with page 863.

Organize *the masses of different locations of the Milky Way by completing the chart below.*

Location	Description of the Mass
within the circle of the Sun's orbit	100 billion times the mass of the Sun
the halo	
	2.6 million times the mass of the Sun

Stars in the Milky Way

Use with pages 864–865.

Describe *the evidence that is used to indicate the extreme age of globular cluster stars.*

_____

_____

Section 30.1 The Milky Way Galaxy (continued)

⸺Main Idea⸺   ⸺Details⸺

Stellar Populations

Use with pages 864–865.

Classify *the populations of stars by completing the table below.*

Classification of Stars	Location of Stars	Brief Description of Elements
Population I Stars		
Population II Stars		

Formation and Evolution of the Milky Way, Maintaining Spiral Arms

Use with page 866.

Sequence *the steps in the evolution of the Milky Way by completing the information below.*

Step One:

Step Two:

Step Three:

Step Four:

Step Five:

COMPARE   Explain how the composition of a star that will form a billion years in the future will differ from the composition of our Sun.

_____

_____

_____

Galaxies and the Universe
Section 30.2 Other Galaxies in the Universe

Main Idea	Details

Details

Predict *Read the title of Section 2. List three topics that might be discussed in this section.*

1. _____

2. _____

3. _____

Review Vocabulary

Use your text to define the following term.

elliptical _____

New Vocabulary

Read the definitions below. Then write the term for each one in the left column.

_____ *is believed to be made up of a form of subatomic particle that intacts only weakly with other matter*

_____ *cluster of galaxies*

_____ *formulaic method of representing the expansion of the universe*

_____ *giant galaxies that emit more radio waves than waves of visible light*

_____ *core of a galaxy where a highly energized object is located*

_____ *star-like objects with emission lines in their spectra*

Academic Vocabulary

Define the following term.

positive _____

_____

Section 30.2 Other Galaxies in the Universe (continued)

Main Idea —————— **Details** ———————————————————

Discovering Other Galaxies

Use with page 868.

Classify *and describe the different galaxies discovered by Hubble. Three of the boxes have been completed for you.*

Type of Galaxy	Description	Subclasses	Subdivisions
			none
		none	none

The Expanding Universe

Use with page 874.

Make *your own simple graph representing Hubble's Law by following the directions given in the Problem-Solving Lab. Label the horizontal axis "Distance", and label the vertical axis "Speed." Include a straight line from the lower-left corner to the upper-right corner.*

Explain *what Hubble's graph shows us about the expanding universe.*

_____

_____

Section 30.2 Other Galaxies in the Universe (continued)

Main Idea ———— **Details** ——————————————————————

Active Galaxies

Use with page 874.

Describe *three unusual properties of radio galaxies.*

1. _____

_____

2. _____

_____

3. _____

_____

Quasars

Use with page 875–877.

Compare and Contrast *two characteristics of stars and two characteristics of quasars in the organizer below.*

Stars:	Quasars:
1. _____ _____	1. _____ _____
2. _____ _____	2. _____ _____

SYNTHESIZE
Hypothesize what the stars at night would look like at night if we lived in an irregular galaxy instead of an elliptical galaxy.

_____

_____

_____

_____

Galaxies and the Universe

Section 30.3 Cosmology

Main Idea	**Details**
	Skim *Section 3 of your text. Look at the headings, photos, illustrations, and captions. Write three questions you have about the information you think may be covered in this section.*
	1. _____
	_____
	2. _____
	_____
	3. _____
	_____
Review Vocabulary	*Use your text to define the following terms.*
radiation	_____
	_____
New Vocabulary	*Use your text to define each term.*
cosmology	_____
Big Bang theory	_____
cosmic background radiation	_____
Academic Vocabulary	*Define the following term.*
correspond	_____

Section 30.3 Cosmology (continued)

Main Idea ———

Details ———————————————————

Models of the Universe

Use with pages 876–877.

Discuss *models of the universe.*

The theory that the universe began as a _____ and has been _____ ever since is called the _____ theory. The Big Bang theory is not an _____ into space, but instead is an _____ of space with _____ going along for the ride.

Supporters of the _____ theory propose that new matter is _____ and added to the universe as it _____ , and thus, the overall _____ of the universe does not change.

Sequence *the steps involved in the explanation of cosmic background radiation. The last step has been done for you.*

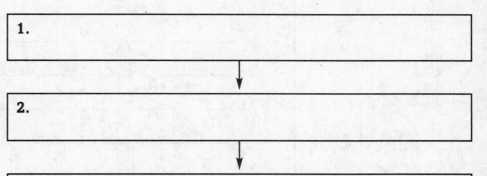

1.

2.

3. The universe expands and cools, and radiation is Doppler shifted to lower energies and longer wavelengths.

State *one of the major reasons why some astronomers do not accept the steady-state universe theory.*

_____

_____

_____

Section 30.3 Cosmology (continued)

Main Idea ━━━━ **Details** ━━━━━━━━━━━━━━━━━━━━━━━━━━━━━━━

The Big Bang Model

Use with page 876.

Identify *the three possible outcomes of a constantly expanding universe.*

Open Universe	Closed Universe	Flat Universe

Explain *how cosmic background radiation helped scientist better understand the origin of the universe.*

_____

_____

_____

_____

_____

┌─ **REAL-WORLD CONNECTION** ─┐
└──────────────────────────────── Hypothesize what the universe might be

like in a flat universe scenario.

_____

_____

Tie-It-All-Together

Compare *The Sun is the center of our solar system, controlling the motions of the planets and other objects. Point out how the universe differs from the Sun in this respect.*

_____

_____

_____

_____

Connect *The Milky Way is a spiral-shaped galaxy. Describe some real-world objects that also have a spiral shape. Include objects that are controlled by a rotational movement.*

_____

_____

_____

_____

Relate *how astronomers are "looking back in time" when they observe light from far away places such as the Andromeda Galaxy.*

_____

_____

_____

_____

_____

Galaxies and the Universe Chapter Wrap-Up

In the "What I Wanted to Find Out" column, copy the questions you listed in the Chapter Preview. In the "What I Learned" column, write down the answers you discovered as you worked through the chapter.

W What I Wanted to Find Out	L What I Learned
1. _____ _____	1. _____ _____
2. _____ _____	2. _____ _____
3. _____ _____	3. _____ _____

Review   *Use this checklist to help you study.*

☐ Study your Science Notebook for this chapter.

☐ Study the definitions of vocabulary words.

☐ Review daily homework assignments.

☐ Reread the chapter and review the tables, graphs, and illustrations.

☐ Review the Section Assessment questions at the end of each section.

☐ Look over the Study Guide at the end of the chapter.

SUMMARIZE

After reading this chapter, list three things you have learned about galaxies and the universe.

_____

_____

_____